Waste Trading among Rich Nations

American and Comparative Environmental Policy
Sheldon Kamieniecki and Michael E. Kraft, editors

Critical Masses: Citizens, Nuclear Weapons Production, and Environmental Destruction in the United States and Russia
Russell J. Dalton, Paula Garb, Nicholas P. Lovrich, John C. Pierce, and John M. Whiteley

Toward Sustainable Communities: Transition and Transformations in Environmental Policy
Daniel A. Mazmanian and Michael E. Kraft, editors

Domestic Sources of International Environmental Policy: Industry, Environmentalists, and U.S. Power
Elizabeth R. DeSombre

Waste Trading among Rich Nations: Building a New Theory of Environmental Regulation
Kate O'Neill

Waste Trading among Rich Nations
Building a New Theory of Environmental Regulation

Kate O'Neill

The MIT Press
Cambridge, Massachusetts
London, England

© 2000 Massachusetts Institute of Technology

All rights reserved. No part of this book may be reproduced in any form or by any electronic or mechanical means (including photocopying, recording, or information storage and retrieval) without permission in writing from the publisher.

This book was set in Sabon by Asco Typesetters, Hong Kong, and was printed and bound in the United States of America.
Printed on recycled paper.

Library of Congress Cataloging-in-Publication Data

O'Neill, Kate, 1968–
 Waste trading among rich nations : building a new theory of environmental regulation / Kate O'Neill.
   p. cm. — (American and comparative environmental policy)
 Includes bibliographical references and index.
 ISBN 0-262-15050-6 (alk. paper) — ISBN 0-262-65052-5 (pbk. : alk. paper)
 1. Hazardous wastes—Management. 2. Hazardous wastes—Transportation. 3. Environmental policy. I. Title. II. Series.
TD1030.O72 2000
363.72'8756—dc21   99-088564

This book is dedicated to my parents, Bob and Sally O'Neill, and to the memory of my grandfather, Donald Frank Burnard

Contents

Series Foreword   ix
Acknowledgments   xiii
List of Tables   xv
List of Abbreviations   xvii

1 Hazardous Waste Trading among OECD Countries: A Comparative Approach   1

2 Out of the Backyard: Hazardous Waste Management as an International Issue   25

3 An Institutional Explanation of the Waste Trade   51

4 Great Britain: Risk Acceptance and the Politics of Flexibility, Diffusion, and Closure   75

5 Germany: Technocracy, Federalism, and Risk Aversion   113

6 The Waste Trade and Environmental Regulation in France, Australia, and Japan   147

7 Conclusion   185

Notes   219
References   261
Index   289

Series Foreword

As we begin the twenty-first century, it is becoming increasingly apparent that many complex relationships exist between domestic and international environmental politics and regulations. Free trade, the formation of economic trading blocs, the globalization of capitalism, rapid advancements in transportation and communications technologies, and other factors are placing mounting pressure on local and global ecosystems. This is forcing nations to address national and international environmental problems and issues simultaneously. Policymakers now face serious obstacles and challenges in the areas of natural resource management and pollution control, within and between national borders. Just as the twentieth century witnessed a sharp rise in economic interdependency between nations, the new century is ushering in greater environmental interdependency between countries. No doubt, these two trends are closely interconnected.

Because threats to environmental quality are more daunting than before, there is a greater need for cooperation between nations than in the past. While the growing body of literature on international environmental politics focuses on the formulation and implementation of international environmental regimes, very little attention is being paid to domestic factors (except when they shed light on a country's position in international negotiations). The emerging literature on comparative environmental policies is also guilty of this omission. As a consequence, and despite present realities and trends, we lack a comprehensive theoretical framework of how domestic factors determine national responses to questions of global environmental degradation and management as well as how they affect the implementation and effectiveness of international environmental agreements.

Fortunately for students of environmental policy, this book provides a major step toward understanding not only the interrelationships between national and international environmental problems, but also how domestic and international environmental politics, governmental institutions, and economic forces intersect. This study focuses on the legal importing of hazardous wastes and analyzes how certain developed nations take on more environmental risk than others because of differences between their styles and structures of environmental regulations. The politics of access to the policy process as well as the politics of regulatory structure, monitoring, and control explain the domestic and international hazardous waste policies of developed countries. These factors, by shaping relations between government, business, and other societal actors who have a stake in the trade of hazardous waste, encourage legal waste importation in some countries and restrict it in others. In more general theoretical terms, the research provides new and important insights into domestic, comparative, and international environmental problems and policy issues, the kinds of which we are likely to face throughout the twenty-first century.

This book illustrates well the kind of works published in the MIT Press series on American and Comparative Environmental Policy. We encourage books that examine a broad range of environmental policy issues. And we are particularly interested in volumes that incorporate interdisciplinary research and focus on the linkages between public policy and environmental problems and issues both within the United States and in cross-national settings. We anticipate that future contributions will analyze the policy dimensions of relationships between humans and the environment from either an empirical or theoretical perspective. At a time when environmental policies are increasingly seen as controversial and new approaches are being implemented widely, the series seeks to assess policy successes and failures, evaluate new institutional arrangements and policy tools, and clarify new directions for environmental policy and politics. These volumes will be written for a wide audience that includes academics, policymakers, environmental scientists and professionals, business and labor leaders, environmental activists, and students concerned with environmental issues. We hope that these books contribute to public understanding of the most important environmental problems, issues, and policies that society now faces and with which it must deal well into the twenty-first century.

Sheldon Kamieniecki, University of Southern California
Michael E. Kraft, University of Wisconsin-Green Bay
Series Editors

Acknowledgments

This book has benefited immensely from the intellectual guidance and support of numerous people over the past few years. In particular, I would like to thank Helen Milner, Hendrik Spruyt, David Vogel, Erika Weinthal, and three anonymous reviewers at MIT Press for their critical input on the manuscript as a whole. Stacy VanDeveer and Alastair Iles generously read and commented on last-minute drafts. Ron Mitchell, Ken Conca, Michel Gelobter, Paul Baldwin, Ellen Feldman, Geoff Taubman, Mark Peranson, Pat Taylor, Jacqui True, Mike Ede, Susan Burgerman, and Michael Turner all provided feedback and/or support at various stages of this work. Finally, this work bears the clear impression of my undergraduate training in political institutions and regulatory economics, for which I can thank Professors Vernon Bogdanor and Peter Sinclair.

Not surprisingly for a book about institutions, I owe much to both formal and informal institutional networks. I would like to thank the Graduate School of Arts and Sciences at Columbia University for its financial support through the Presidents Fellowship Program, the Managing the Atom Project at the Belfer Center for Science and International Affairs, Harvard University for postdoctoral support (in particular, Steve Miller and John Holdren), and the Department of Environmental Science, Policy, and Management, University of California at Berkeley. The Smith Richardson Foundation supported my research in the United Kingdom, and I would like to thank agencies in Germany and Australia for assisting my inquiries. The Environmental Studies Section of the International Studies Association helped out with some travel money, but, most important, it has provided a supportive intellectual community for young scholars in this field. The same can certainly be said for the Gep-Ed e-mail list. Jennifer Clapp, Joe Jupille, Jonathan Krueger, and Laura Strohm, who have worked or continue to work on waste trade issues, have been extremely generous with data and help over the years. Ambuj Sager provided the most creative title suggestion. I would also like to thank Clay Morgan and Melissa Vaughn at MIT Press. Special thanks to Wil Burns for compiling the index, and to Tracey Brieger for help with the proofs.

Family and friends, nearby and far away, have been unstinting in their support and friendship, keeping my sense of humor alive and picking up the pieces when necessary. Finally, my deepest gratitude goes to my parents, Bob and Sally O'Neill, and my sister, Jenny, for their unfailing support and understanding of my work. This book is also dedicated to the memory of my grandfather, Donald Frank Burnard, whose love of Australian landscape and nature inspired all of his grandchildren.

Tables

Table 2.1 Some illustrative examples of hazardous wastes
Table 2.2 Generation of hazardous waste by country, OECD members, 1994 and 1997
Table 2.3 Summary of transfrontier movements of hazardous wastes, select OECD countries, 1989–1993
Table 3.1 Summary of hypotheses related to environmental regulation
Table 4.1 Imports of hazardous wastes to the United Kingdom, 1988/9–1993/4
Table 4.2 Imports of waste to the United Kingdom, 1995 and 1996
Table 4.3 Net imports (exports) of hazardous wastes for selected OECD countries, 1989, ranked by net import level
Table 4.4 Net imports (exports) of hazardous wastes for selected OECD countries, 1991, in rank order
Table 4.5 Imports of hazardous waste to England and Wales by country of origin, 1993–94
Table 4.6 Waste disposal methods in the United Kingdom, early 1990s
Table 4.7 Disposal licenses in England and Wales, 1991–4
Table 4.8 Data available on hazardous waste disposal capacities, Great Britain, 1984
Table 5.1 Summary of transfrontier movements of wastes, Federal Republic of Germany, 1989–1991
Table 5.2 German special waste exports, 1988–1993

Table 5.3 Transfrontier movements of waste, Federal Republic of Germany, 1996
Table 6.1 Net imports (exports) of hazardous wastes by Australia, France, and Japan, 1989–1993
Table 6.2 Main variables and findings
Table 7.1 Contending explanations
Table 7.2 Main variables and findings

Abbreviations

ALP	Australian Labor Party
BATNEEC	Best Available Techniques Not Entailing Excessive Costs
BMU	Bundesministerium für Umwelt, Naturschutz und Reaktorsicherheit
BPEO	Best Practicable Environmental Option
BPM	Best Practicable Means
BSE	Bovine Spongiform Encephalitis
CDU/CSU	Christian Democrat/Christian Socialist Parties
CFCs	Chlorofluorocarbons
COPA	Control of Pollution Act, 1974 (U.K.)
DDR	Deutscher Demokratischer Republik
DETR	Department of Environment, Trade, and the Regions
DGXI	Directorate General XI
DoE	Department of the Environment
EA	Environment Agency
EC	European Community
ECJ	European Court of Justice
EM	Ecological Modernization
EMAS	Environmental Management and Auditing Scheme
ENDS	Environmental Data Service
EPA	Environmental Protection Agency
ESA	Environmental Services Association
EU	European Union

FDP	Free Democrats Party
FoE	Friends of the Earth
FRG	Federal Republic of Germany
GNP	Gross National Product
HMIP	Her Majesty's Inspectorate of Pollution
HWI	Hazardous Waste Inspectorate
ICC	International Chamber of Commerce
IEP	International Environmental Politics
IGAE	Intergovernmental Agreement on the Environment
IPC	Integrated Pollution Control
IPPC	Integrated Pollution Prevention and Control
ISO	International Standards Organization
LDC	Less Developed Country
LDP	Liberal Democratic Party
MAFF	Ministry of Agriculture, Fisheries, and Food
MITI	Ministry of International Trade and Industry
NAFTA	North American Free Trade Agreement
NAWDC	National Association of Waste Disposal Contractors
NGO	Nongovernmental Organization
NIMBY	Not in My Back Yard
NRA	National Rivers Authority
NSW	New South Wales
OECD	Organization for Economic Cooperation and Development
PCBs	Polychlorinated biphenyls
PR	Proportional Representation
SOEnD	Scottish Office of Environment and Development
SPD	Social Democratic Party (Germany)
T&D	Treatment and Disposal
UBA	Umweltbundesamt
UNCED	United Nations Conference on Environment and Development
UNDP	United Nations Development Programme

UNEP	United Nations Environment Programme
USAID	United States Agency for International Development
WCA	Waste Collection Authority
WDA	Waste Disposal Authority
WMI	Waste Management International
WRA	Waste Regulation Authority

1
Hazardous Waste Trading among OECD Countries: A Comparative Approach

The international trade in hazardous waste has thrived since the 1970s. The Organization for Economic Cooperation and Development once estimated, famously, that "in 1984 ... on average, a consignment of hazardous wastes crossed an OECD frontier every five minutes, 24 hours per day, 365 days per year" (OECD 1993b:9). Officially, about 1 million tonnes, or 5–10 percent of hazardous wastes produced by the rich countries is legally traded.[1] In reality the figure is probably much higher. This trade is most commonly associated with egregious cases of waste dumping by firms from the United States and Western Europe on poorer countries in Africa, Latin America, or the Caribbean. However, more than 80 percent of the trade is transacted between the industrialized countries of the world and is legal by domestic and international standards. Such trading is the most visible symptom of the overall crisis facing waste management across industrialized countries, and it represents a problem that has been given far too little consideration by either policymakers or scholars concerned with managing the trade from the international level. This work deals with a specific empirical puzzle: why some OECD countries voluntarily import such wastes in the absence of pressing economic need. Great Britain, despite its size, population density, and the fact that it is a democracy, is one of the world's largest legal importers of hazardous wastes. Germany, despite its reputation for environmental responsibility both at home and abroad, is a major waste exporter.

The broader question of why states differ in their responses to similar international problems calls for an examination of countries' national level characteristics. I argue that the most important factors determining whether an advanced industrialized democracy imports waste are features of its system of environmental regulation. In the field of comparative politics, institutional explanations of political outcomes focus on the way interactions between governments and other societal actors are mediated by institutional structures and practices.[2] Institutions are also the focus of the growing literature on international environmental politics, where the institutions in question are the rules and structures established under international environmental agreements and the extent to which these elicit compliance from signatory states. To date there have been few attempts to build bridges between these two literatures, for example, to examine how different national responses to environmental problems shape international environmental and political outcomes (e.g., routes and direction of trade in hazardous wastes). I address these issues in this book and argue that national differences in regulatory systems are key in understanding why some wealthy countries take on the risk of importing wastes and others do not.

An institutional approach helps us understand and compare different policy and real-world political outcomes where the preferences of government, industry, and other societal actors are broadly similar across countries, and where differences between national wealth and state capacity are minimal. I develop here a qualitative model of environmental regulation that can be applied, with minimal adaptation, across countries. The model is grounded in the notion that regulatory systems are best seen as "nested": hazardous waste and waste trade regulations are part of broader systems of environmental regulation, whose procedural aspects in turn are driven by broader characteristics of national political systems. Institutions mediate and constrain the preferences of different actors, or stakeholders, in the system, thereby altering political outcomes: they are not deterministic, nor are they malleable representations of actor preferences.

The cases studied in this volume are Britain, Germany, France, Australia, and Japan. As outlined below, some of the findings challenge conventional wisdom: for example, Britain and France, usually considered to be centralized and therefore strong states, are shown here to be neither, at least in the area of environmental policy. Finally, a common problem cited in both international and comparative studies of environmental politics is that of measuring and comparing effectiveness of environmental policies. I overcome this problem by identifying a measurable dependent variable—the net imports of hazardous wastes—which in turn serves as a proxy for the amount of risk to its population a country's government and firms are willing to take on.

Spilling across Boundaries: Global Hazardous Waste Problems

Whether one sees wastes as a problem of social control or a market opportunity depends on one's own point of view (Thompson 1994:201).[3] Hazardous, or toxic, waste generation continues to increase around the world. According to OECD figures, average annual hazardous waste generation among its member states increased by roughly 65 million tonnes in the early 1990s alone (OECD 1997a). Much of this waste is highly toxic—containing very long-lived substances such as PCBs, dioxins, and asbestos—and is generated for the most part by industry, though household, agricultural, and medical practices contribute to the problem. Hazardous waste management has no big disaster, no equivalent of Chernobyl or the discovery of the hole in the ozone layer, to galvanize action. Rather its crises have built up slowly, affecting large numbers of people over a long period of time, as in the case of Love Canal in northern New York State, or "Minamata Syndrome" in Japan, where in the 1950s firms dumped mercury in local fishing grounds, poisoning tens of thousands of people across several generations. Public awareness of potential threats from waste generation and inadequate disposal, along with a virtual absence of public trust in the abilities and motives of government actors, has led to high levels of opposition to the siting of any new waste disposal facilities in all but the poorest neighborhoods and areas. In turn, the economic cost of complying with disposal regulations has left industries scrambling for cheaper solutions for dealing with their stockpiles of wastes.

Under these circumstances, the emergence of an international trade in hazardous wastes can hardly be surprising, and for a long while the dumping of wastes on less developed or poorer countries was left unchecked. In the late-1980s, however, the outcry over these practices led to the UNEP-sponsored Basel Convention on the Transfrontier

Movement of Hazardous Wastes and Their Disposal (1989), which aims to ban all trade in waste from developed to less developed countries.[4] The European Union, too, regulates the movement of wastes within and out of its territory, official bodies such as the OECD Secretariat in Paris monitor the trade and collect relevant data, and international nongovernmental organizations such as Greenpeace carry out important roles as international watchdogs of the trade.

Despite this activity, the trade continues. The dependent variable of this study is a country's propensity to legally import hazardous wastes, measured in terms of its annual net imports of hazardous wastes from abroad. Determining why countries import wastes is a harder question than determining why they export them. It touches on a range of issues as to how government actors interact with the waste disposal industry and broader societal groups and the degree of control (if any) that any one group exercises over the trade; it also raises many broader questions of state-society relations in industrialized democracies. Answering this question might also provide important clues as to how to regulate the trade and prevent industrializing countries from making the same mistakes in the future.

The Argument: National Regulatory Systems and Waste Importation

The OECD countries have the most advanced regulatory systems in the world, and studies have shown that their overall success rate in dealing with pollution at the national level does not vary significantly. Therefore the concern here is with the procedural aspects of countries' regulatory systems—"the way things are done"—rather than the substantive policy goals and outcomes in each case. To that end, the key differences among countries with respect to the waste trade lie in two features of their regulatory systems. First, countries differ in terms of their structures of hazardous waste management and regulation—the allocation of responsibilities among different agencies and levels of government as well as the structure of the hazardous waste disposal industry. The second set of differences concerns national styles of environmental regulation, which mediate government-industry-society (or stakeholder) relations in the making and implementation of environmental policy. In this study I use two variables—access to the policy process and mode of policy implementation—as indicators of regulatory style. In each case these rules and practices mediate the preferences of the actors involved, producing very different outcomes across the five OECD countries examined here—Britain, Germany, France, Australia, and Japan—despite quite similar configurations of actor preferences.

Regulatory Structure: Coordinating Waste Importation and Management in Britain and Germany

National regulatory structures may be classified on a continuum between fully centralized and diffuse. A diffuse structure in this context is one where regulatory responsibilities are divided among many different agencies at both the national and local levels and where the waste disposal industry is privately owned and highly competitive. A fully centralized regulatory structure is, in contrast, one where a single government agency holds all waste management responsibilities and where the waste disposal industry is a public monopoly. Regulatory structure matters in that it affects the ability of government officials to monitor, control, and coordinate the actions of its citizens: the less coordination there is in the system, the more leeway firms have to import hazardous wastes. Following principal-agent theory, therefore, the more decentralized a system is, the less likely are the principals (national government officials) to be able to monitor and affect the behavior of the agents (firms importing hazardous wastes).

In Britain regulatory responsibilities for hazardous waste management are devolved to scores of local authorities, which in turn issue waste importation permits. These authorities oversee a highly competitive, privately owned waste disposal industry, and at the same time have been locked in an almost continual battle with the central government over their responsibilities, resources, and political legitimacy in the absence of a clearly codified set of shared understandings about their role. These factors have meant that there is little effective monitoring of waste importation practices at any point in this chain of command, and that there is also a great disparity in the incentive structures facing both regulators and firms. In comparison, Germany has a mixed public-private disposal industry—with high levels of government intervention in determining its practices, costs, and goals—and a federal system. The sixteen Länder and the federal government cooperate over policy; their constitutionally defined roles, responsibilities, and sources of funding help simplify the monitoring problem and help bring actor preferences and incentives into line.

Regulatory Style: Access and Implementation
Two variables in particular are indicative of a country's regulatory style: access to the policy process and mode of policy implementation. In systems where access to the policy process is more open, no one societal group is especially privileged with respect to others in the process of policy planning and formulation. A system is relatively closed when government officials, or government officials working in tandem with business interests, comprise the sole policy community of importance, and broader societal interests are excluded. In particular, public opinion is often excluded from debates. Britain has a closed policy process, where industrial interests (waste disposal firms) are privileged over broader societal and environmental groups (Greenpeace, Friends of the Earth, etc.). In contrast, Germany has a much more open system. This is demonstrated by examining their respective policy communities and wider political opportunity structures (following Kitschelt 1986).

The policy community is the group of government officials and representatives of outside groups who determines the agenda, goals, and content of environmental policy. In Britain this group is very narrow, consisting of bureaucrats and industry representatives, and operates under conditions of relative secrecy. In Germany the process is more public. It also includes (and this is one of the defining features of German environmental policy) groups of outside economic, scientific, and legal experts, whose input into the process has led to a very precautionary, or anticipatory, approach and has broadened the scope of regulatory goals.

Identifying the political opportunity structure involves looking at broader institutional facets of these systems, such as a country's electoral system and its processes of judicial review. For example, Britain has a first-past-the-post electoral system whereby the political party with the most seats wins leadership of parliament and of the country.

This system virtually precludes effective representation at the national level of parties that cannot claim extensive territorial bases of support. Germany's semiproportional system has been instrumental in allowing the rise of the Green Party, whose influence has been both direct and indirect in the ways it has forced the mainstream parties to incorporate environmental concerns into their agendas.

The ways in which environmental policies are implemented also affects the waste disposal industry's freedom of action. German firms have to adhere to strict, nationally applied regulatory standards, based usually on the precautionary principle, which mandates action in the face of risk to the environment. Their British counterparts work according to the "best practicable means" criterion, where each firm is obliged to protect the environment as best it can given the means at its disposal. This, along with the more reactive principles governing British environmental policy, means that firms have much more leeway in choosing courses of action that have an impact on the environment.

In sum, the chapters that follow show that Britain is more likely than Germany to import wastes because its regulatory system—diffuse, closed, and flexible—allows industry much more freedom of action, precludes societal interests that oppose waste importation, and has, in fact, proven a major obstacle to government efforts to implement change. The government has yet to implement the ban on waste imports it announced in 1991. Germany displays a very different configuration of government-industry-society relations. Greater input of environmental and technocratic interests into the policy process and higher levels of government control of the waste industry have led to what can only be termed a highly risk-averse approach to the hazardous waste trade. This system has also led to the highest waste disposal costs in Europe—hence explaining both the absence of countries willing to ship their wastes to Germany and Germany's own high levels of exports.

Other Cases: France, Australia, and Japan

This theoretical framework can be applied to other industrialized democracies, here France, Australia, and Japan. France is similar to Britain in terms of the basic parameters of its regulatory system. It is also a large—in fact, the largest—legal net importer of hazardous wastes. Its structure of environmental regulation in particular is highly conducive to waste importation. As with Great Britain, responsibilities are allocated to a wide range of regional authorities ill-equipped to adequately control the problem. With respect to the access variable, it is also closed: the French state is not highly permeable to societal interests in the way that the German or Australian states are.

Australia is an interesting case for several reasons. Given its size and available space, one would expect it to be a waste importer. It is not, however, and in fact it has until recently been a waste exporter. Another important characteristic is its highly centralized waste management structure. Although Australia differs qualitatively from Germany in several respects (for example, a different type of federal structure prevails in Australia), broadly the two match up in terms of regulatory structures and styles. Japan, finally, is the outlier in this study. Although its regulatory system suggests that it would be a waste importer, it is not. In my analysis of the Japanese case I therefore examine conditions external to the model that have led to such an outcome, paying particular attention to the lasting effect of the "Big Four" pollution cases of the 1950s and how these continue to influence the actions of official and industry actors.

Methods and Alternative Explanations

The central claim of this work—that institutional factors can significantly constrain the actions and choices of "free" political actors—is by no means uncontested in the field of political science.[5] To establish this claim, it is necessary to evaluate the regulatory argument on its own terms via an in-depth analysis of institutional structures, stakeholder preferences, and policy outcomes in each country, then compare it with alternative explanations. The dependent variable is measured using data on transfrontier waste movements supplied by the OECD and national governments (where available). The choice of OECD countries controls for level of development and for the existence of advanced regulatory systems. There is also a fairly consistent set of data available for the relevant variables. Most importantly, OECD countries display significant variation in terms of their respective styles and structures of environmental regulation.

International factors do not provide sufficient explanatory leverage here. An argument based on trade and other linkages between the OECD member states as facilitating factors does not capture the specific direction that the legal trade takes. A regime-based argument does not work either: the Basel Convention was implemented too late to affect the actions of these states, and in any case governs primarily waste exports from OECD to non-OECD countries. I identify two alternative domestic-level explanations for waste importation practices. The first is based on the incentives faced by state actors in allowing the import or export of hazardous waste—for example, governments allow waste importation because the benefits reaped far outweigh attendant risks to the environment. The second, based on the notion of comparative advantage, argues that legally traded wastes are exported to countries with superior disposal facilities (and/or a high waste disposal capacity)—for example, specialist incinerators that can safely immobilize highly toxic substances.

Related Issues and Broader Questions

This approach demonstrates the important role played by institutional constraints in the domestic environmental policy process with respect to an international issue—the waste trade. It touches on some of the broader research questions driving contemporary scholarly work on the politics of the international environment: the links between domestic and international politics with respect to the environment, crossnational comparisons of risk management techniques and risk propensities, and the implications of polity structure for environmental regulation.

The trade in hazardous wastes among wealthy countries can be studied from the perspectives of comparative politics, international relations theory, and trade theory. In a very important sense, this work concerns the linkages between domestic and international environmental politics, linkages that have to date been underspecified in the relevant literature. A focus on domestic political practices and constraints is especially important in light of debates around the "internationalization of environmental protection" (Schreurs and Economy 1997). Interaction between international and domestic politics is increasing in this field through the implementation of international environmental commitments, the role of transnational political actors, the intersection of trade and security debates with environmental concerns, and the ways domestic environmental politics themselves are being transformed through engagement in the international system. This work concerns itself at different levels with all of these, focusing in particular on how different national approaches to environmental politics have affected transboundary transmission of pollutants and states' ability or inability to fulfill international environmental commitments. The concluding chapter examines how international political ideas and processes have started to transform the traditional landscape of environmental policy in the last decade of the twentieth century and the several important ways the model presented intersects with processes and theories of regulatory change in a changing international political context.

In some ways, the focus of the book on hazardous waste trading and institutional theory appears narrow. Broader political economy and political culture arguments occasionally raise their heads, as in the case of Thatcherite Britain, and are dealt with on that basis. A very different book could be written on the ways these forces, should they be easily identifiable, shape regulatory and institutional politics. However, by focusing on a narrower set of questions, I hope to generate an agenda that can in the future be widened and extended to different public policy issues, and help illuminate broader questions of international and comparative political economy.

Domesticating the International: The Field of International Environmental Politics

How do states respond to common environmental threats? Phrased at this general level, this question captures the common thread that has run through recent work in the fields of international relations and comparative politics on environmental problems. Yet to date little work has been done toward capturing interactive dynamics between practices at the national level, environmental degradation at the global level, international cooperative responses in the form of regulatory agreements, and subsequent domestic-level policy shifts.

Most countries are facing very similar sets of environmental problems, which may be considered local (e.g., deforestation, overexploitation of natural resources such as fossil fuels, biodiversity protection, etc.), transnational (e.g., transboundary air and water pollution), or global (pollution of the global commons: the oceans and the atmosphere). Over time, but especially over the last twenty-five years, states have evolved distinctive national responses to these problems in the form of environmental regulations and rules governing the behavior of private and public actors engaged in polluting activity. At the same time, there has been an increased recognition of the need for states to coordinate their activities to overcome transboundary environmental issues and the degradation of the global commons. (Both sets of developments date more or less from the first UN conference on the environment held in Stockholm in 1972.[6]) Coordination has taken the form of international regulatory agreements, which states then implement via changes in their own regulatory policies, targets, and goals.[7] The system of global governance that has emerged from this process demonstrates a level of embeddedness and maturity that few had predicted at the outset (Haas 1999), despite some weaknesses evidenced in, for example, global climate change negotiations.

The dominant research agenda for scholars of the environment in the field of international environmental politics (IEP) has been driven by central questions in the field of international relations theory: Why do states cooperate in an anarchic international environment in the absence of a sovereign authority, and how do they manage the constraints and opportunities existing at the level of the international system? For many, the study of environmental cooperation above all contributes to understanding interstate dynamics in the context of a "new" issue area. The environmental field has, however, considerably extended the insights of the cooperation literature in mainstream international relations theory by enhancing understanding of the role of nonstate actors in international politics and compliance with and implementation of international regulatory regimes and institutions.[8] In particular, work on nonstate actors has opened the field of international relations to the possibility of actors with different bases of power than traditional state actors—for example, the knowledge-based authority of scientific and other expert groups, the economic power of organizations such as the World Bank, and the moral authority and watchdog status of many NGOs.[9] Work on implementation and compliance is, at last, pushing the boundaries of the field's research beyond the outdated propositions of Realism—that state cooperation itself is a puzzle—and opening up the "black box" of domestic politics in interesting ways (Simmons 1998; Weiss and Jacobson 1998). Some interesting work is being done, for instance, on issues of scale and global environmental politics, especially when problems have cross-scale implications (Cash and Moser 1998; Moser 1998).

However, the field's focus on states, international organizations, and the need to design international agreements primarily to discourage cheating has led to a "top-down" and in many ways partial understanding of international environmental issues, especially as they intersect with domestic political norms and practices.[10] Domestic-level factors are on the whole included only as they determine national interests at the bargaining table (Sprinz and Vaahtoranta 1994).

The emerging focus of this literature is on how to evaluate the effectiveness, or success, of environmental regimes, a project that has implications for international regimes in other issue areas (Sprinz 1996; Bernauer 1995) and marks a departure from earlier work on regime formation.[11] Two factors in particular account for this shift, both related to a realization that the achievement of lasting cooperation among states is not an end in itself. First, the pressing nature of environmental problems has led to growing demands for adequate solutions. At the same time, doubts are emerging about the extent to which existing agreements actually remedy international environmental degradation. The second factor relates to the particular nature of these agreements. International environmental agreements are regulatory agreements, part of a subset, albeit a large one, of the set of international agreements. Unlike human rights or traditional trade agreements, which tend to target government actors directly, regulatory agreements are implemented indirectly. They require signatory governments to enact rules and regulations at the domestic level to change the behavior of private actors responsible for the problem in the first place (Mitchell 1994:13). Understanding how these work and what obstacles may exist in each country helps us to design agreements and predict their likely effectiveness. This applies not only to environmental agreements but also to other types of international regulatory agreements, which include financial regulations and, increasingly, nuclear weapons monitoring and international trade agreements.

The term "institutional effectiveness" is used in many different ways throughout the literature. It seems intuitively apparent that the best measure of regime effectiveness is the extent to which the agreements under examination actually contribute to the improvement of the physical environment. However, there are several other ways to define and measure the notion of effectiveness in practice—Oran Young (1994), for example, identifies no fewer than six distinct dimensions of regime effectiveness.[12] Most commonly, the conception used is state *compliance*: the degree to which states alter their behavior to comply with the terms of an agreement.[13] To date, most studies of institutional effectiveness have focused on factors located primarily (although not always wholly) at the international level. These include the nature of the negotiation process, the design of institutional rules and the extent to which these affect state behavior, and the role of nonstate actors at the international level.

Studies that focus explicitly on the negotiation process itself argue that it contains built-in flaws, which in turn inhibit the overall effectiveness of resulting agreements. Building from detailed analyses of interstate negotiations within individual issue areas,[14] authors evaluate and offer prescriptions for improving the convention-protocol method, the dominant mode of international environmental negotiation (Sand 1990; Susskind 1994).[15]

Other authors investigate the design of formal and informal institutional rules and the degree to which these change actor incentives to evoke compliance (Chayes and Chayes 1991, 1993; Birnie and Boyle 1992; Mitchell 1994; Weiss and Jacobson 1998). Haas, Keohane, and Levy 1993 and Keohane and Levy 1996 seek to show how international institutions augment effectiveness at all stages of the policy process (agenda setting, policy formation, and policy implementation). A similar top-down approach is taken by Victor, Raustiala, and Skolnikoff (1998), who examine how systems for implementation review at the international level and participation in the international regime affect agreement implementation at the national level. Finally, many authors examine the role played by nonstate actors at the international level in shaping both the regime-building process and the behavior of states.[16] As Karen Litfin argues, "while the nation-state system may remain at center stage, it is clearly being pulled simultaneously in many directions.... [T]he knowledge-based nature of environmental problems has opened up the playing field to a profusion of unconventional actors" (Litfin 1993:111).

The tendency for this research to be driven almost exclusively by the wider questions of the international relations agenda has led to some clearly identifiable problems for studies of IEP. First, these works tend to ignore the political factors that may cause or contribute to international environmental degradation in the first place. Instead, questions of causality are left to natural scientists or economists. This omission is particularly striking, as it seems that political scientists should be well equipped to deal with such issues. As Conca and Lipschutz argue, different international relations approaches to the environment "share a tendency to treat [environmental] change as originating outside the social systems in which its consequences are assessed" (Conca and Lipschutz 1993:6). They lay the blame for this on over reliance on the existing tool kit of international relations theory, namely, "intergovernmental agreements and international regimes, debates about partially yielding up sovereignty, arguments about collective action, and so on" (ibid.:7).

Second, certain problems have arisen with respect to the regime effectiveness literature. For example, scholars have found it extremely difficult to derive and consistently maintain an operational definition of the concept of effectiveness, one that can easily be transformed into an empirically applicable measurement.[17] Definitions of effectiveness abound, and there is often a gap between authors' initial definition of the term and the way they subsequently use it. Without good, recent data on actual environmental changes, and with the implicit bias toward statecentric analysis, most of the literature ultimately relies on the notion of state compliance, the degree to which states alter their behavior to comply with the terms of an agreement.[18]

The arguments I present here seek to move beyond the question of why states cooperate under anarchy by asking instead what political factors located at the domestic, or national, level of analysis determine distinctive national responses to issues of international environmental change and management. So far such factors have been used in this literature primarily to explain national interests toward regime formation.[19] However, they also play an important role in understanding to a greater depth the processes and actors at the domestic level that may constrain or facilitate the process of regime implementation and, more broadly, the fight against environmental degradation. They are absolutely crucial in understanding different national responses to a national environmental problem with transnational and international implications.

Domestic politics are starting to move to center stage in a number of recent works in the field. Preeminent among these are Schreurs and Economy 1997, Raustiala 1997, Weiss and Jacobson 1998, Weinthal 1998, and Baldwin 1999. All emphasize both the difficulty of and the need for qualitative crossnational studies in understanding how national and international environmental politics interact. To understand the interplay between national and international actors, between regulatory systems and the global environment, it is ultimately necessary to draw on different disciplines or fields for theoretical insight. The most important of these is the field of comparative environmental policy, which provides the basis for theorizing about the key differences between regulatory systems.

Internationalizing the Domestic: Comparative Environmental Policy

A number of influential works comparing environmental policies across countries have been published in the last two decades, asking questions such as: why different management and regulatory practices have evolved crossnationally with respect to the environment, what determines "policy effectiveness" in this area, and what this tells us about their political systems (Lundqvist 1978).[20] Most of this literature focuses on the environmental policy and politics of the advanced industrialized nations, whose systems are advanced enough and display enough similarity (especially in terms of specified goals and concerns) to make crossnational comparison feasible.[21] Furthermore, while some of this literature examines countries' environmental politics as a whole, others compare national behavior across individual issue areas, or, alternatively, compare policy practices in the environmental sectors with other key areas of public policy.[22] The study of comparative environmental policy has never really taken root in the theoretical mainstream of the field of comparative politics. Rather it has its origins in, and draws from, the literature on new social movements and comparative public policy.

Recent literature in this field has addressed several questions, or dependent variables. First and foremost is the question of relative levels of effectiveness, namely, why some countries are more successful than others in reducing levels of environmental degradation.[23] One result of this literature, aside from some of the problems of actually measuring the impact of a given set of rules, is that the observed variation between many of the main countries used in these comparative cases is surprisingly low, despite the differences between them (Jasanoff 1991). This in turn has launched a body of work on the relative efficiency or cost-effectiveness of different systems (e.g., Moe and Caldwell 1994) and on the "implementation deficit"—the gap between the goals of environmental policy and its outcome.[24] A related set of studies examines the factors determining national capacity for achieving environmental goals (Jänicke and Weidner 1997; Jänicke and Weidner 1995). Different environmental impact assessment procedures across countries and choice of policy instruments have also generated fruitful research.[25]

Analysts have taken several different approaches in studying these questions.[26] Society-based approaches have been driven by the emergence of environmental movements across different countries, a phenomenon that in turn drove theoretical work on New Social Movements. In particular, such works have examined political opportunities and constraints confronting national movements, the available formal and informal forms of organization and mobilization, and the framing processes that mediate between opportunity and action (McAdam, McCarthy, and Zald 1996:2). Some, such as the work of Ronald Inglehart on postmaterialist values and work by Riley Dunlap and others on the 1991 Gallup Health of the Planet Survey, examine the role of public opinion in environmental politics.[27] State-centric approaches in the environmental field often focus on less developed or nondemocratic countries (Peluso 1993; Ribot 1993; Desai 1998). However, studies of government regulatory failure in the light of environmental problems (Eckersley 1996), national interest formation (Sprinz and Vaahtoranta 1994; Raustiala 1997), and bureaucratic or organizational politics (Weale 1992; Weale et al. 1996) also fit under this rubric. Most works in the field study only two or three cases in depth, though some take a more quantitative approach, comparing indicators across most, if not all, OECD countries—an approach that suffers from some superficiality in its conclusions but allows for a good "first cut" on comparative policy.[28]

This book centers on the significance of institutional variables for the understanding of crossnational differences in environmental policies related to the global environment. Institutions may be defined as "identifiable practices consisting of recognized roles linked by clusters of rules or conventions governing relations among the occupants of these roles" (Young 1989:5). As Weale puts it, "the argument of those who favor the institutional approach is that public policies need to be understood in the light of the specific configurations of institutions and organizations that exist within the political system" (Weale 1992a:52).

Earlier institutional work in this field analyzed domestic political systems and cultures to shed light on different national environmental performances (e.g., Lundqvist 1980; Knoepfel and Weidner 1986; Reich 1984) or on political systems as a whole (Enloe 1975). Some of the problems associated with this sort of analysis, chiefly "an inevitable superficiality in their accounts of national political cultures" (Buller, Lowe, and Flynn 1993:179), led to analyses of "the steady institutionalization of environmental concerns" (ibid.:180), and national policy styles. Examples of works in this field include Vogel 1986, Aguilar 1993, Forester and Skinner 1987, Moe and Caldwell 1994, Héritier et al. 1994, and Brickman, Jasanoff, and Ilgen 1985.[29] More recent scholarship explicitly attempts to identify and trace the linkages between different levels of institutional influence on environmental policies and outcomes (Ikenberry 1988). Hence, another primary concern of this literature is explaining the ways in which certain institutional forms affect specific patterns of pollution control policy. For example, while Moe and Caldwell (1994) argue that these practices are primarily a function of a country's governance system (presidential or parliamentary), Vogel (1986) makes an historical argument, tracing how the evolution of government-business relations in the U.S. and the U.K. with respect to environmental policies differed over time.[30]

Much of this literature situates environmental problems firmly within national borders, without considering interaction effects between states or how environmental problems spill across national borders.[31] This focus is changing, especially with the growth in (top-down) influence of transnational organizations (such as the EU) and processes on national policymaking.[32] However, few of these studies attempt to relate domestic differences in environmental politics and policy to state interaction in the context of an international environmental issue. This work is intended to redress this imbalance. It also overcomes some of the problems the comparative field shares with its international counterpart. By examining national differences using a single, measurable indicator —hazardous waste importation—I avoid problems of identifying, measuring, and comparing policy effectiveness or success across the environmental sphere.

The Pollution Haven Hypothesis and Economic Approaches to the Global Environment

Some of the trade-environment and international economics literature also sheds light on domestic-international interactions and effects. Economists argue that differences—both quantitative and qualitative—between national systems of environmental regulation have important implications for the conduct of economic activity in an increasingly internationalized world economy. These differences matter particularly in determining the movement of goods and services across national boundaries. This argument works at two levels. First, regulatory differences create "distortions" in the movement of goods and services across national frontiers—an insight that can also be applied to the movement of pollutants across borders. Second, as a result of increased interdependence, regulatory systems will themselves begin to change, moving toward some form of international harmonization (or rapprochement) of regulatory standards.

With respect to the first level of the argument, a growing body of literature concerns the environmental effects of trade and trade liberalization and foreign direct investment.[33] One of the chief hypotheses of this literature is the "pollution haven" hypothesis, which in its strong form posits that "countries would intentionally integrate lenient environmental regulations into their industrial policies in order to attract manufacturing away from countries that demanded environmental quality" (Strohm 1993:135), a process also known as "regulatory arbitrage" (Leyshon 1992). Various empirical studies and reviews of the literature have argued that in fact, all else equal, environmental regulations play very little role in the location decisions of multinationals compared with other factors such as labor costs (Dicken 1992). There are two weaker forms of the hypothesis that could be applied instead, derived from Copeland's classification of international environmental externalities.[34] First, countries with weaker systems of regulation are more likely to produce pollutants that cross national boundaries and are less likely to take effective action to remedy the problem. Second, while companies might not be willing to incur the expense of relocation in order to avoid environmental regulations, they would be likely to relocate at least part of their activity (e.g., waste disposal) to a country with less stringent regulations. I argue against the pollution haven hypothesis about weaker, or more lax, regulatory systems. The cases here show that procedural aspects of the most advanced regulatory systems facilitate or prohibit environmentally risky importation.

Literature addressing the second level of the argument involves the question of international regulatory harmonization as a result of increased international openness. Many of these studies are concerned with the direction of harmonization within regional economic blocs to the lowest or highest common denominator, with serious implications for the international environment (Vogel 1995). My argument corresponds more closely to arguments made by some political scientists and sociologists: that national political styles and practices remain resistant to external pressures for change (following Weale 1995; Hollingsworth, Schmitter, and Streeck 1994; Buller, Lowe, and Flynn 1993). This point is made in more depth in chapter 3. A more sophisticated analysis of regulatory changes as a result of interdependence is provided by Jacobs

(1994), who analyzes the concept of regulatory rapprochement, a more complex phenomenon that occurs at several levels. The first is regulatory harmonization—"the standardization of regulations in identical form." However, less fundamentally, regulatory rapprochement also includes "mutual recognition, or the acceptance of regulatory diversity as meeting common goals," and "coordination or the gradual narrowing of relevant differences between regulatory systems, often based on voluntary international codes of practice" (Jacobs 1994:32). Jacobs argues that governments can adapt to the pressures of global economic and environmental interdependence by adopting elements of all three types of regulatory rapprochement, and that this is already happening—in the European Union, for example.

Cross-national Risk Management: The Salience of the Hazardous Waste Trade

The politics of the environment are in essence the politics of risk and uncertainty, much of which remains unquantifiable and the subject of controversy across all branches of environmental politics. Issues of hazardous waste disposal and management occupy a prime position in discussions over how best to determine, compare, and act on levels of risk posed to human beings and the ecosystems in which they live.[35]

Risk enters this project at several levels. At a very basic level, the generation and disposal of hazardous wastes pose clear risks to human health and to local ecosystems. As the volumes of hazardous wastes generated have increased exponentially over recent decades, and as industrial processes threaten to release ever more toxic substances into the environment, these issues have been thrown into sharp relief with the emergence of the international waste trade. More than this, however, debates in most advanced countries over hazardous waste facility siting and the cleaning of older contaminated sites has shown a vast divide between popular perceptions and "expert assessments" of these risks.[36] The resulting "Not In My Backyard" (or NIMBY) phenomenon has, in combination with extremely high costs of waste disposal in many countries, driven many firms to seek alternative disposal routes. In many cases, this has led to the exportation of wastes to countries willing, for a variety of reasons, to receive them (Hilz 1992). In the face of these (actual and perceived) environmental risks and the increased political and economic costs associated with adequate waste disposal, it is not hard to explain waste exportation. A more difficult phenomenon to explain is why some countries, in particular highly industrialized democracies with vocal environmental movements, should choose to allow waste imports.

This work speaks to the body of literature that examines differences in national risk propensities, assessment procedures, and attitudes.[37] Despite the continued salience of debates over risk perception, trust, and communication in the North American and West European environmental political arenas, few studies attempt to establish why some countries are more risk-averse than others with respect to international environmental issues.[38] My argument here is that institutional constraints play an important role in determining national risk propensities, as exhibited in policy outcomes. For example, while both British and German public opinion exhibit considerable (and similar) levels of concern over hazardous waste disposal, the German system has built into it many more access channels whereby public opinion is expressed in the policy process. Thus certain biases built into the British political system have meant that British policy outcomes regarding waste importation reflect a more risk acceptant attitude than other countries. Similar analyses apply to France, which parallels Great Britain, and to Australia, whose administrative structure is most similar to Germany's among the cases examined here.

Political Structure and Environmental Management
A final set of questions concerns the relationship between domestic political structures and environmental management, a long-standing issue in the field of environmental politics and indeed one of the broader issues raised in comparative policy analysis more generally. Many scholars have examined, for example, the relationship between democracy (or regime type) and environmental politics, and much has been written on the emergence of environmentalism as a new social movement.[39] In the global arena, many of these debates are now taking place in the context of discussions about sustainable development and its relationship to political and economic practices.

The countries examined in the following chapters are all democracies. However, despite certain broad similarities, they all display different styles and structures of environmental policy and management. These differences are exhibited in many ways that defy standard dichotomies such as that between "strong" and "weak" states, or even between "pluralist" and "corporatist" modes of interest mediation. In a sense, this approach of challenging common conceptions of state strength across issue areas follows that of Krasner (1978). It also challenges some of the findings made in the political economy literature about strong and weak industrialized states (e.g., Katzenstein 1978). As many authors have pointed out, environmental problems pose distinct sets of challenges to "politics as usual," challenges that can rarely, if ever, be resolved through the development and application of "technocratic fixes" (Weale 1992a). The above brief discussion of the politics of risk in the hazardous waste debate illustrates this point.

The opportunity to examine the systems of environmental regulation across a number of countries—Britain, Germany, France, Australia, and Japan—reveals a fascinating variety of regulatory styles and structures and, within those, distinctive national approaches toward defining environmental issues, incorporating societal interest groups, dividing regulatory responsibilities, and applying policies. For example, hazardous waste management in the context of the trade is a complex, multilevel issue with both global and extremely local implications. Most of the countries I examine define waste management as a local issue, and in some the power to issue waste importation permits—which would seem to be a national government responsibility—is delegated to local authorities or governments. In turn, whether these governments are situated within a unitary or a federal polity also affects their ability to control the trade. This study, somewhat counterintuitively, finds that seemingly more decentralized federal systems appear to be better at managing and coordinating hazardous waste trade and management than are unitary systems.

Countries also vary greatly in the ways in which they incorporate different societal interests into the policy process, either within the relatively narrow confines of the immediate policy community or in the broader context of electoral representation of views or judicial review of legislation. These variations include not only who is represented in the process—be it industrial interests, expert representatives, and/or environmental groups—but also the extent to which broad or narrow channels of access to the policy process are institutionalized or protected within a political system.

Overview of the Project

In this first chapter I have outlined the main themes and implications of my arguments regarding hazardous waste trading at the intersection of international and comparative environmental policy. In chapter 2 I discuss hazardous waste management as a global issue, focusing on the waste trade and the international regime that has emerged in response to the problem, in particular the Basel Convention and measures taken by the EU to restrict transfrontier movements of wastes. I also set the scene for subsequent case analysis, critiquing the international regime on the grounds that it ignores the problems of waste management in developed countries, instead focusing primarily on the transboundary aspects of the trade. In chapter 3 I discuss the theoretical approach in more depth, identifying and drawing out the logic behind the main hypotheses regarding regulatory structures and styles and examining the hypotheses derived from the two main alternative explanations. I also discuss case selection and conclude with two potential obstacles the empirical analysis needs to overcome: how to account for issues of regulatory change in this context and the problems of gathering and using data on hazardous waste trading and management. Chapters 4 through 6 are empirical. Chapters 4 and 5 deal with Britain and Germany respectively, and chapter 6 extends the analysis to the regulatory systems of France, Australia, and Japan.

In the concluding chapter I pull together the empirical analysis and assess my findings against the hypotheses identified in chapter 3. I address some of the wider theoretical and policy implications of my work, going back to the broader questions discussed above, and examine how a regulatory approach can be generalized across global environmental issue areas. I argue that a greater understanding of factors determining national responses to questions of global environmental degradation and management is both a worthy goal in itself and may also aid in the task of operationalizing and measuring the effectiveness of international environmental regimes. I also discuss the potential for, and driving forces behind, medium- and long-term change in countries' waste management policies and their overall strategies of environmental regulation; finally I assess the future of the international hazardous waste trade management regime.

2
Out of the Backyard: Hazardous Waste Management as an International Issue

Countries that accept imports of hazardous wastes from abroad are essentially taking on environmental risks that other countries, for a variety of reasons, are not willing to absorb. That wealthy, industrialized democracies willingly take on these risks, in the apparent absence of any form of coercion, flies in the face of much of the conventional wisdom about state or government behavior and runs counter to many of the assumption of mainstream international relations theory in both its realist and neoliberal forms. It also runs contrary to received wisdom that public opinion in many of these countries runs strongly against waste importation and, given that most of the countries under analysis are democracies, that these preferences would be translated into public policy. In this chapter I trace the emergence of the global waste trade and the international regulatory regime, discuss some of the key conflicts and identify the main actors in the field, thus setting up the context for later analysis. I argue that the way the waste trade is framed—as a transboundary issue—overlooks the root of the problem: the growing incapacity of most countries to adequately handle the wastes they generate. Hazardous wastes are all too often dealt with at the local level, where their effects are most often felt. However, waste management has long since become a problem with not only local but also national and international implications.

Hazardous Waste Regulation in Industrialized Countries

Hazardous waste management and regulation are among the most contentious of all environmental issues for government officials to deal with, but at the same time one of the most important. Most industrialized countries have had comprehensive waste management statutes on the books since the early-1970s, yet few would claim they had successfully controlled the problem.[1] In this section I describe the context in which the waste trade emerged by exploring some of the common problems facing domestic hazardous waste regulators—including issues of waste definition, identification, and classification, and problems posed by environmental damage, market failures, and public opinion—and the evolution of the waste disposal industry.

What Is a Hazardous Waste? Classifications and Management Practices

Wastes are the unwanted byproducts of industrial or household activity (be it production or consumption). They exist in many forms; however, what actually constitutes a "hazardous waste" is hard to pin down and is, in practice, highly context-specific.[2] A waste in one context may in another be a useful and valued commodity, a simple example being plastic shopping bags, seen for the most part as throw-away items in many Western countries but highly prized and continually reused elsewhere in the world. What constitutes a hazard also varies with the context. For example, wastes considered not particularly hazardous in industrialized countries—such as sewage—are among the greatest causes of disease and ill health in many less developed countries.[3] A common definition that must be rejected at the outset is that wastes are substances with no economic value. If that were the case, then there would be no trade in hazardous wastes. Furthermore, many wastes contain materials that can be recovered and reused: this tension underpins current controversies surrounding international waste trade negotiations (Krueger 1999).

Even in industrialized countries, delineating the problem boundaries of this issue area for the purposes of policy analysis is not easy. As Brian Wynne argues, these "have little to do with objective, natural dictates, and much to do with social factors, pragmatic necessities, and administrative purposes" (Wynne 1987:49). Official definitions, classifications, and reporting requirements of hazardous wastes vary extensively from country to country.[4] For example, while some countries require reporting of all wastes, regardless of whether they are destined for recycling or for final disposal, others require only that wastes destined for final disposal be reported. Another practice is the use of different terms such as "hazardous," "toxic," "difficult," or "special" wastes. This has led to a major problem for international regulation and control, and the establishment of a common international scheme for the definition and classification of hazardous wastes is one of the primary points of controversy in Basel Convention negotiations (see below).

In general, "[m]ost countries have used a definition based on three criteria. These are particular types of hazardous wastes, industrial processes from which the waste is defined as hazardous; and/or substances, the presence of which is indicative of a potential hazard" (Forester and Skinner 1987:29). In practice many countries have simply drawn up lists of substances considered to be hazardous. This is problematic because the quality of testing processes in many countries has not kept pace with the exponential increase of new chemicals on the market (World Resources Institute 1987; Wynne 1987). The OECD defines hazardous wastes as "wastes which, if improperly managed or disposed, could harm man [sic] and/or the environment because they are toxic, corrosive, explosive, combustible etc." (OECD 1993b:7).[5]

Wastes deemed hazardous are often categorized into high-, intermediate-, and low-risk (Asante-Duah, Saccomano, and Shortreed 1992), depending on the degree of risk posed to the environment and the types of management practices required.[6] They range from dioxins and heavy metals (such as mercury, cadmium, or lead), highly hazardous even at fairly low levels of concentration, to many organic wastes, generally considered to be less hazardous. Hazardous wastes result from a huge variety of production practices and may take many forms, from barrels of liquid waste to discarded rubber tires and batteries to light ash. Table 2.1 lists some examples of hazardous wastes. These wastes usually require specialized disposal techniques such as processing, incineration, and protected landfill—techniques that are expensive and require extensive monitoring. Waste disposal technologies are often listed in descending order of desirability (and descending order of costs), generating the well-known waste management hierarchy, a tool used as the

Table 2.1
Examples of hazardous wastes

Sector	Source	Hazardous waste
Commerce, services and agriculture	Vehicle servicing	Waste oils
	Airports	Oils, hydraulic
	Dry cleaning	Fluids, etc.
	Electrical	Halogenated solvents
	Transformers	PCBs
	Hospitals	Pathogenic/infectious wastes
	Farms/municipal parks, etc.	Unused pesticides, "empty" containers
Small-scale industry	Metal treating	Acids, heavy metals
	Photofinishing	Solvents, acids, silver
	Textile processing	Cadmium, mineral acids
	Printing	Solvents, ink, and dyes
	Leather tanning	Solvents, chromium
Large-scale industry	Bauxite processing	Red muds
	Oil-refining	Spent catalysts, oily wastes
	Chemical/pharmaceutical manufactures	Tarry residues, solvents
	Chlorine production	Mercury

Source: Batstone, Smith, and Wilson 1989.

basis of many national waste management strategies.[7] Following this approach, the main techniques are:

- Policies of waste reduction at source.
- Recycling, which involves the recovery of both energy and materials.[8]
- Chemical treatment or incineration to destroy, convert, or immobilize hazardous constituents.[9]
- Disposal on land, either directly or in a lined or unlined landfill. Historically, as the U.S. experience shows, landfill has been the most popular method of waste disposal. However, it has become increasingly unpopular as the problem of dealing with old and abandoned or contaminated sites has become apparent.[10]
- Disposal of wastes at sea, through dumping or shipboard incineration. Although many countries used to engage in these practices (Forester and Skinner 1987:53), they are now largely banned by national policy and by international agreement.

Prevailing characterizations of hazardous wastes exclude the majority of municipal (or household) wastes, instead covering mainly industrial and agricultural wastes. Nuclear wastes are excluded from this analysis as they pose very different regulatory, political, and structural properties, reflected in national and international governance regimes.[11] Their importance for national security and a much more centralized industry structure are but two of these properties. Finally, although many wastes cross national boundaries through natural dispersion processes, as is the case with acid rain in Europe, international and national regulation of hazardous wastes focuses on solid or liquid wastes, that is, wastes that require physical transportation to their site of disposal.

Issues in Waste Regulation: Shared Problems and Challenges

Environmental Damage and Environmental Justice
There are several ways in which hazardous wastes pose a threat to the environment.[12] At the most basic level, waste production is an environmental threat through sheer accumulation.[13] The production of hazardous wastes and the development of new hazardous substances, particularly in industrialized countries, has increased exponentially over the past decades, and in the absence of credible commitments to waste reduction policies on the part of governments and firms, these trends are likely to continue. Table 2.2 shows an increase of more than 65 million tonnes worldwide between the 1994 and 1997 reporting dates.[14] Careless or accidental mismanagement of hazardous wastes has afflicted every industrial country. Love Canal, the Maquiladora situation on the U.S.-Mexican border, and the case of well contamination in Woburn, Massachusetts, that provided the basis of Jonathan Harr's bestselling book, *A Civil Action* (Harr 1995) are crises that continue to resonate with the U.S. public; other countries have their own equivalents.

Improper handling and disposal of hazardous wastes impose additional burdens on the environment. They affect human health through leakage of toxins into ground water, the soil, or the atmosphere, and local habitats and ecosystems may be damaged or even destroyed as a result of waste disposal.[15] Thus waste products are causal agents in many forms of pollution. Their effects may be immediate, occurring on

Table 2.2
Generation of hazardous waste by country, OECD members, 1994 and 1997

	Year reported (published 1994)	Hazardous waste generation (1,000 t.)	Year reported (published 1997)	Hazardous waste generation (1,000 t.)
Australia	n.d.	300	1992	426
Austria	1991	620	1995	915
Belgium	n.d.	27,000	1994	27,530
Canada	1990	6,080	1991	5,896
Denmark	n.d.	112	1993	91
Finland	1987	250	1992	367
France	1992	7,000	1992	7,000
Germany	1990	6,000	1993	9,020
Greece	1990	423	1992	450
Ireland	n.d.	66	n.d.	66
Italy	1991	3,246	1991	3,387
Japan	n.d.	666	n.d.	666
Netherlands	n.d.	1,500	1993	2,600
New Zealand	1982	60	199-	110
Norway	1990	200	1991	220
Portugal	1987	1,043	1994	1,365
Spain	1987	1,708	1987	1,708
Sweden	1985	500	1985	500
Switzerland	1991	736	1993	837
Turkey	n.d.	300	1989	300
United Kingdom	1991	2,956	1993	1,957
United States	1989	197,500	1993	258,000
Total		258,266		323,411

Source: OECD 1994a, table 2; OECD 1997a, table 2. Figure for Belgium counts all industrial wastes produced in Wallonia only. Only wastes destined for final disposal need to be reported in Britain and Germany. Netherlands hazardous waste generation includes 845,000 tonnes of contaminated soil. n.d.: no data provided. The difference between the waste generation figures for the United States and Europe arises largely because the United States defines large quantities of dilute wastewaters as hazardous wastes, while in Europe these materials are managed under water protection regulation.

exposure to the waste, and/or long term, as the waste decomposes or leaches into the surrounding environment and enters the food chain (Batstone, Smith, and Wilson 1989:3). Furthermore, unlike air pollution for example, environmental degradation from hazardous wastes disposal is likely to be very localized in its effects and hence much more concentrated. There is no one-to-one correlation between the disposal of hazardous wastes and damage to the environment. However, the risk of environmental damage increases with the level of hazard posed by the waste and as disposal facilities become more crowded and less capable of handling the wastes processed through them. A recent study using data from Belgium, Denmark, France, Italy, and the U.K. found that women who live within three kilometers of hazardous waste landfill sites have a 33 percent higher risk of having babies with nonchromosomal birth defects than those living further away. This study was highly publicized when was released, and led to calls for further investigation and more systematic monitoring of the situation.[16]

Environmental damage caused by the incorrect disposal of hazardous wastes has important political and economic ramifications over and above the direct cost to the environment. Cleaning up old and contaminated waste sites imposes a substantial economic cost on regulatory authorities, and the political fallout from the discovery of such sites can be considerable (Strohm 1993:6). The most ambitious national program to tackle the problem of old sites is Superfund, a U.S. federal government program that is widely considered an expensive failure.[17] Most other countries have yet to come to terms with their legacy of old and abandoned sites.

At the domestic level, issues of waste disposal are of high salience to local communities. Being both visible and highly localized in their effects, waste disposal sites have often proven to be a focal point for conflict between communities, corporations, and government agencies, as the U.S. experience has shown. Above all, hazardous waste management as part of the much larger class of toxics issues—including air pollution, dangerous industrial and construction practices, and many other activities—is inseparable from the agenda of the environmental justice movement. Hazardous waste disposal facilities are often located in minority and poor communities—communities made up of people who are least able, for economic and political reasons, to assert the tenets of NIMBYism, and who may, for economic reasons, either "choose" to endanger themselves or be forced to live in a toxic area where the only affordable housing is located. The practice of siting disposal facilities in such communities has come under strong and consistent attack from activist groups. However, it remains a contentious issue (in the U.S. more so than in the countries studied here), and the parallels with the hazardous waste trade to LDCs are strong.[18]

Environmental Risk and Risk Perceptions
The risks posed by hazardous wastes have generated crises in the development of disposal technology, the siting of facilities, and the complex interrelationships between regulators, elected representatives, and communities. The "traditional" science of risk assessment has failed to assuage public mistrust of the efficacy of existing waste management practices, and the siting of new waste disposal facilities has become extremely difficult.[19] Countries all over the world are having severe trouble in this regard (Yakowitz 1993). Reliance on the "science" of quantitative dose-response and exposure assessment, which involves strict adherence to the principles of technical rationality, has proven to be problematic. Instead of quelling public fears, the use of these techniques often reveals "a wide gulf between the scientists' assessment of the risks associated with a proposed product or project and the public's" (Armour 1993:185).[20] EPA studies within the U.S. have consistently shown that "public and professional rankings of environmental hazards are markedly divergent" (McKenzie 1994:92, citing EPA 1987). According to one analysis, "it would appear that, in general, people assess the risks associated with living near a hazardous waste treatment facility as being so great that virtually no reasonable amount of compensation, by itself, can have much impact" (Portney 1988:60).[21]

Observations of these differences in public and official risk assessments, and the social construction of risks, raise a number of concerns and conceptual issues. At a very basic level, writers as diverse as Ulrich Beck and Rachel Carson point to what is almost an underlying malaise of technoindustrial society: the extent to which we have come to rely on processes and materials that bear the seeds of potential catastrophe.[22]

In particular, the uneasy combination of unpredictability, toxicity (or radioactivity), and possible disaster presented by modern practices has never been adequately encompassed by or resolved in any institutional context.[23] Hazardous waste-related concerns are central to these discourses. On a more prosaic level, these debates are about much more than risk. They are also about trust (in government regulators, for instance) and about communication between state and societal actors via the channels through which they are mediated (La Porte and Metlay 1996; Entwistle 1997; Kasperson, Golding, and Kasperson 1998). As the essays in Munton 1996 argue, the NIMBY syndrome is better viewed through the lenses of communication and trust than that of the irrational fears of an ignorant public.[24]

These debates recur crossnationally, and a theme apparent in recent discussion of regulatory change is that traditional, top-down modes of regulation are no longer suitable for the sorts of problems regulators of toxic and radioactive substances face. In times of crisis and national controversy issues of risk, trust, and communication come into sharp relief. Many studies have explored the BSE ("Mad Cow Disease") debacle in the U.K. and the contrasting debates in Europe and the U.S. over the introduction of genetically modified organisms (GMOs) into consumption and agricultural production cycles in the late-1990s.[25] The problems posed by hazardous waste management are more "day-to-day" than these, and usually much less high-profile. This absence of actual crises makes it difficult to achieve consensus about regulatory reform.

Market Failures

The nature of offsite waste disposal and the involvement of third parties —waste disposal companies or brokers to whom waste-generating firms sell their wastes—exacerbate pollution externalities and other market failures associated with many forms of industrial production. Externalities are the unintended "side effects" of production or consumption that pose a social cost (positive or negative) on actors outside the firm. They are not reflected in the profits generated by the firm or, therefore, in decisions about output (Helm and Pearce 1990).

As Brian Wynne points out, there is an important difference between the service provided by waste disposal firms and other sorts of services:

the result is invisible to the purchaser, who cares only that the wastes be removed from his or her care.[26] This creates a different set of incentives than those governing transactions in normal goods. Exporting firms or communities have little self-interest in ensuring that the wastes arrive at their ultimate destination, and importing firms, communities, and indeed countries have a financial incentive to import more wastes than are socially desirable (Wynne 1989:123). Monitoring becomes extremely complex in these situations, but all the more important as the number of affected parties increases. Another serious set of problems has to do with monitoring the large number of waste generators, especially the small firms that are more likely to seek offsite disposal. As the waste disposal industry becomes more multinational, too, it becomes harder for national agencies to hold individual firms to account, or even to identify responsible ownership.

Government Response: Looking for an Exit Route
As Wynne argues, a point underlined by the above arguments, "[i]n most countries, hazardous waste management is in a state of internal flux and public strife. International controls are crippled by problems and divergences in national approaches" (Wynne 1987:25). Waste disposal facilities in most industrialized countries are becoming increasingly congested, few new ones are being built, and yet few measures are being undertaken by firms or by national regulatory authorities to enact effective waste reduction measures at the point of production. These patterns are being repeated across Europe (Hilz and Ehrenfeld 1991). According to Yakowitz, "all [OECD] countries are now experiencing difficulties ranging from moderate to extremely severe in trying to site new waste management facilities. New hazardous waste sites are almost impossible to obtain in Western Europe and Japan. Australia is also having great difficulty in this respect" (Yakowitz 1993:137–8).

In most industrialized countries, the cost of waste disposal has vastly increased over the past few years, due to more stringent environmental regulations and more advanced disposal technologies, particularly the movement away from landfill methods to more advanced incineration and processing.[27] At the same time, costs of international waste transportation have fallen (Strohm 1993:133). It is therefore often cheaper for companies to export their wastes to countries where disposal is less costly, and to pay agents in those countries to dispose of the wastes. Another motive for exporting wastes is the decline in disposal capacity in most industrialized countries. As landfill sites become inoperable, and the siting of hazardous waste disposal facilities becomes more contentious, many companies feel they have no option but to export their wastes to countries more willing to receive them. These factors form the focus of the literature on waste exportation.[28] If one accepts these conclusions about the crisis facing domestic waste management systems, the interesting question is what determines a country's willingness to import hazardous wastes, especially as it is the industrialized, congested countries that undertake the bulk of waste importing.

Globalization of Waste Management: The Trade and Related Issues

Under these circumstances, the emergence of the international waste trade is not surprising. The first officially recognized transboundary shipments of hazardous wastes occurred in the 1970s, and since then there has been a substantial growth in the trade.[29] Estimates vary, but one 1985 study puts the rate of increase in hazardous waste exports for the United States and the European Community at 2–4 percent per year (OECD 1985, cited in Hilz 1992:38). By far, the greatest part of the waste trade is legal, carried out with the consent of the relevant authorities in the importing and exporting countries and in compliance with international regulations. However, a significant proportion of the trade is illegal, occurring without the explicit consent from importing authorities or in violation of international agreements.[30]

Beyond the waste trade, there are at least three additional ways in which waste management has become an international issue. First, the major waste disposal firms have widened their global reach: most own and operate disposal plants outside their country of origin. For example, Générale des Eaux, the largest French waste disposal company, is currently the largest European provider of waste disposal services and the fifth largest global provider, with installations as far-flung as Chile, Australia, and New Zealand. Mergers of large waste disposal firms are almost monthly news.[31] Second, many multinational corporations dispose of their wastes within their countries of operation. This practice has led to many problems: even if companies attempt to maintain codes corresponding to those held in their country of origin, the local infrastructure may not support these standards, nor might there be adequate communication between the company and the local authorities.[32] Third, wastes are often discharged (from land) or dumped (from a ship) into the marine environment. Technically, ocean dumping has been banned since the 1990 strengthening of the 1972 London Dumping Convention. However, no international regulations govern land-based discharge (Blowers 1996:178–9).

The emergence of the waste trade can therefore be traced to the expansion of hazardous waste production in industrialized countries over the last few decades, and to the increased difficulty and cost of siting and building disposal facilities as old ones were filled or became obsolete (Hilz 1992; Wynne 1987; see table 2.2). The growth of the waste trade was also greatly facilitated by trade liberalization under the General Agreement on Tariffs and Trade (GATT) and the European Community. The attendant sixtyfold increase in world trade since 1950 (Vogel 1995:11) opened national borders for international movements of environmentally undesirable goods and services. In terms of volumes of wastes traded, exact figures are hard to come by. The OECD has estimated that "in 1984 ... on average, a consignment of hazardous wastes crossed an OECD frontier every five minutes, 24 hours per day, 365 days per year" (OECD 1993b:9). According to Montgomery 1995, "the UNEP estimates that the Western European countries annually trade 700,000 tons of hazardous wastes among themselves, and that the U.S.A. and Canada each export 200,000 tons, primarily to each other. Moreover, until the new ban's implementation, European countries legally exported about 120,000 tons of hazardous waste to developing countries every year" (4).

These figures represent roughly 5–10 percent of wastes produced by OECD countries (Strohm 1993:132).[33] Table 2.3 summarizes transfrontier movements of hazardous wastes in select OECD countries, reported in 1994 and 1997. The waste trade is also big business for the waste disposal industry, which has become increasingly privatized and globalized in recent years (Crooks 1993; Cooke and Chapple 1996), and is extremely profitable: imported waste often generated more than a thousand dollars per tonne in revenue. Greenpeace lists over 300 firms involved in the waste trade alone (Vallette and Spalding 1990, appendix A).

According to the OECD, roughly 80 percent of the legal waste trade occurs between OECD countries, while 10–15 percent goes to Eastern Europe and the remainder to less developed countries (see also Montgomery 1995; Hilz 1992). The Third World Network has identified three main trading routes: from industrialized countries to Great Britain, from industrialized countries to Africa, and from the United States to Canada and Mexico (Third World Network 1989:41). Two other major routes have recently been recognized from Western Europe to countries in Eastern Europe and the former Soviet Union—both before and after the Cold War—and from industrialized countries to Southeast Asia (Puckett 1994). Two legal forms of international waste trading are given less attention here. The first are transports of waste among the United States, Canada, and Mexico under NAFTA: a report issued by the Commission for Environmental Cooperation states that in 1995 close to 27,000 tonnes of hazardous wastes were shipped to Mexico from the United States (O'Neill 1999a). The second, also involving the United States, are shipments of wastes to, and dumping on, Native American communities and lands. These cases have become part of the environmental justice literature but deserve much more attention as valid cases of international trade, given the semisovereign, or domestic dependent nation, status of Indian reservations.[34]

Following the blaze of publicity given to cases of illegal waste dumping, many less developed countries, particularly in Africa, have, with the support of international environmental organizations such as Greenpeace, begun to refuse waste shipments from the industrialized world.[35] In fact, the number of (documented) proposed waste schemes now far outweigh those that go ahead (Vallette and Spalding 1990; Montgomery 1994, 1995): according to Montgomery, this is a more interesting issue than why hazardous waste shipments are proposed in the first place.[36] This study asks the converse question: Why do some industrialized countries continue to accept wastes from abroad, despite the associated problems and risk?

Table 2.3
Summary of transfrontier movements of hazardous wastes, select OECD countries, 1989–1993 (in tonnes)

	1989			1990		
	Imports	Exports	Net	Imports	Exports	Net
Australia	0	500	(500)	0	1,000	(1,000)
Austria	50,981	86,773	(35,792)	19,180	68,162	(48,982)
Canada	150,000	101,083	48,917	143,811	137,818	5993
Denmark	26,842	8,978	17,864	16,376	9,214	7162
France	n.d.	n.d.	n.d.	458,128	10,552	447,576
Germany	45,312	990,933	(945,621)	62,636	522,063	(459,427)
Italy	0	10,800	(10,800)	0	19,968	(19,968)
Japan	5,125	40	5,085	397	0	397
Netherlands	88,400	188,250	(99,850)	199,015	195,377	3638
New Zealand	0	200	(200)	0	0	0
Norway	0	8,078	(8,078)	0	16,532	(16,532)
Portugal	n.d.	n.d.	n.d.	0	1,954	(1,954)
Spain	27,413	280	27,133	82,269	20,213	62,056
Sweden	33,863	45,012	(11,149)	47,223	42,636	4,587
Switzerland	7,684	108,345	(100,661)	6,688	121,420	(114,732)
United Kingdom	40,740	0	40,070	34,983	0	34,983

Source: OECD 1997a, table 1.
Notes:
Net exports are in parentheses.
Due to differences in national definitions of hazardous wastes, great caution should be exercised when using these figures.
U.S. data was excluded due to absence of data on imports: the U.S. requires written notice and consent for exports only.
Australian data only concerns permits for final disposal.
Austria enforced its new ordinance on hazardous wastes in 1991.
Canada enforced new legislation on transfrontier movements of hazardous wastes in 1992.

Table 2.3
(continued)

1991			1992			1993		
Imports	Exports	Net	Imports	Exports	Net	Imports	Exports	Net
0	3,200	(3,200)	0	275	(275)	0	0	0
111,595	82,129	29,466	79,107	70,023	9084	28,330	83,998	(55,668)
135,161	223,079	(87,918)	123,998	174,682	(50,684)	173,416	229,648	(56,232)
15,200	21,758	(6,558)	100,244	15,858	84,386	n.d.	n.d.	n.d.
636,647	21,126	615,421	512,150	32,309	479,841	324,538	78,935	245,603
141,660	396,667	(255,007)	76,375	548,355	(471,980)	78,219	433,744	(355,525)
0	13,018	(13,018)	n.d.	21,627	n.d.	n.d.	19,365	n.d.
397	0	397	n.d.	n.d.	n.d.	n.d.	n.d.	n.d.
107,251	189,707	(82,456)	250,355	172,906	77,449	236,673	163,180	73,493
0	21	(21)	0	208	(208)	n.d.	10,469	n.d.
2,415	14,636	(12,221)	64,070	14,545	49,525	81,207	16,639	64,568
1,147	292	855	5,638	457	5,181	7,195	815	6,380
81,597	6,578	75,019	66,356	15,803	50,553	104,716	13,943	90,773
34,195	63,801	(29,606)	61,725	22,185	39,540	82,933	22,484	60,449
6,416	126,564	(120,148)	10,471	132,138	(121,667)	8,360	125,840	(117,480)
46,714	525	46,189	44,673	0	44,673	66,294	0	66,294

Decreases in German data between 1989 and 1990 are due to German unification in 1990.

Dutch data excludes imports and exports of nonferrous metals destined for recycling.

New Zealand data before 1993 includes PCB exports only; exports of hazardous waste going to recovery only.

The increase of imports to Norway is due to huge amounts of aluminum salt slag being sent there for recovery.

Portugal enforced new legislation on transfrontier movements in 1992.

Spain changed its regulations concerning hazardous wastes between 1989 and 1990.

For the U.K. and Germany, only wastes going to final disposal have to be reported.

International Responses to the Waste Trade: The Basel Convention and EC/EU Directives

Pressure to Cooperate

The growing internationalization of hazardous waste management has in turn exacerbated many of the problems faced by domestic regulators. For example, the risk of environmental degradation across national frontiers is arguably higher than if wastes are processed closer to home, especially when one takes risks of accident in transit into account. The OECD has found that "given the cost differentials for various disposal methodologies, hazardous waste which crosses frontiers destined for disposal in another country is likely to be waste considered *highly hazardous*" (McNeill 1985:9; emphasis in original).[37] Risk perceptions are amplified when wastes are of foreign origin. Moreover, waste importation is seen by many countries, and not only LDCs, as a violation of national sovereignty, an insult added to injury, as the tone of recent media coverage in Britain of waste imports makes clear.[38] Finally, the absence of an international sovereign authority and of viable mechanisms for assigning liability and arranging for compensation, or even for monitoring waste transactions and volumes, made cooperation among states in this issue desirable.[39] Such cooperation was made necessary by the intense media scrutiny of the social inequities of waste dumping on poorer and less developed countries, seen as environmental colonialism in its worst and most obvious form.[40]

Pressures for regulation of the waste trade between industrialized and less developed countries began to increase in the late-1980s in the wake of several well-publicized attempts (both successful and otherwise) to offload hazardous materials at various ports throughout the Caribbean and Africa. One of the most notorious cases was that of the Khian Sea, which, carrying a load of municipal ash from Philadelphia (labeled as fertilizer), attempted to put into port at several Caribbean countries before succeeding in dumping part of the ash in Haiti. After Haitian authorities became aware of what the barrels actually contained, the ship was sent away. It subsequently traveled to five continents, undergoing several name changes, before showing up in 1988, its holds empty, having dumped its contents somewhere in the Indian Ocean

(Vallette and Spalding 1990). In December 1994 Greenpeace demanded that the U.S. government repatriate the roughly 2000 tonnes of ash still in Haiti.[41]

The evolution of the international waste trade regime has been piecemeal. In its early days, agreements actually conflicted in terms of their underlying principles and norms, with some seeking to monitor and restrict the trade, others to facilitate it, and some to ban it.[42] Following pressure from LDCs and international NGOs, existing agreements are now converging on the principle that waste trading from richer to poorer nations should be banned, although full consensus over which countries should be allowed to import wastes, which substances should be regulated, and how the ban should be enforced remains out of reach.

The main—but by no means the only—international agreement governing the waste trade is the UNEP-sponsored Basel Convention on the Transboundary Movement of Hazardous Wastes and Their Disposal, signed in 1989 and in effect by 1992. In 1991, in protest of the Basel Convention, twelve African countries signed the Bamako Convention, which bans the importation of hazardous wastes from any country, and since then many other less developed countries have also banned waste imports. The Bamako Convention came into effect on March 20, 1996. The 1991 Lomé IV Convention banned trade between EC member states and former colonies in Asia, the Caribbean, and the Pacific. North American waste trading is now regulated under the terms of the NAFTA agreement. Historically, U.S. firms operating in the border economic zone—*maquiladoras*—have been obliged to ship their wastes back across the border. This mandate, always dubiously applied, could be lifted in 2000 (O'Neill 1999a). Other states that have implemented bilateral waste trade agreements facilitating trade include France and Germany. Several regional agreements exist as well, such as the Izmir Protocol of the Barcelona Convention (1996), governing waste trade in the Mediterranean region (Cubel-Sánchez 1997), and the Waigani Convention (1995), part of which addresses waste dumping in the South Pacific (Hyman 1997). The EU is the most important force in regulating trade among industrialized countries, and a discussion of the early evolution of its waste trading policies follows analysis of the Basel

Convention. The OECD also plays a role in monitoring and recording the waste trade among its members and outside the organization.[43]

The Basel Convention: Evolution of a Controversial International Regime

The Basel Convention was signed by 22 countries in Basel, Switzerland, in March 1989, and subsequently signed by a total of 118 nations (Puckett 1994:54).[44] The regime rests on the principle of prior informed consent, whereby potential exporters have to receive written consent from government officials in the importing nation before the shipment can occur. Furthermore, the onus is placed on the exporter to ensure that the waste will be disposed of in an "environmentally sound manner," a phrase poorly defined in the original Convention. Under the original terms of the convention, exportation of wastes should only occur if the exporting country does not have the facilities to dispose of the wastes properly or if the wastes are to be used as raw materials by the importing nation. The agreement also established that the UNEP would set up a facility for the gathering and distribution of data and information regarding the waste trade. By May 1992 the convention had been ratified by the required number of states, and thus entered into effect. Notable exceptions to treaty ratification include many less developed countries and the United States, which claimed that its domestic regulations were adequate to ensure the proper control of hazardous waste movements.[45]

Many flaws and points of contention emerged from the 1989 meeting.[46] Critics of the 1989 Convention range from LDCs and NGOs (such as Greenpeace) to powerful multinational corporations. One of the major criticisms leveled at the Convention in its original form was that it allowed for major loopholes, through the recycling and raw materials provisions (Puckett 1994; Third World Network 1989; Hackett 1990; Vallette and Spalding 1990). Second, opponents claimed the Convention monitored only the actual transfer of wastes from one country to another, and therefore cannot ensure that wastes are properly disposed of at their final destination (Wynne 1989; Laurence and Wynne 1989; Gourlay 1992). Third, it did not ensure that transparency and flows of information will increase as a result of the agreement: as yet, the UNEP facility for monitoring the waste trade has not been established. Finally, under the terms of the convention, there are no provisions for assigning liability or assuring compensation for environmental damage across national frontiers. The agreement has no enforcement capability and provides no protection against impecunious governments that are willing to import wastes at the possible expense of their populations (Birnie and Boyle 1992:341–3).

Subsequent Conferences of the Parties addressed the "recycling loophole" and strengthened the core principles of the treaty: in 1994 parties agreed voluntarily to ban all exports of wastes, both for disposal and for recycling purposes, from OECD members to nonmembers. However, controversies remained. At the 4th Meeting of the Parties, in Malaysia in February 1998, delegates made only limited progress on drawing up comprehensive lists of hazardous wastes for the purposes of the Convention. They disagreed on which countries should be allowed to belong to Annex VII (the group allowed to continue importing wastes: long-standing OECD members and Liechtenstein) and on the extent and availability of bilateral agreements between Annex VII and non-Annex VII countries to continue trading in hazardous wastes. These disagreements have yet to be resolved, and could lead to the Basel Convention being challenged under World Trade Organization rules as a trade restriction (Krueger 1999). Furthermore, the ban amendment had by early-1999 been ratified by less than a quarter of the parties required for it to come into force. It is likely to be a long time before this happens. While the major players in the waste industry support restrictions on illegal dumping of wastes (a reputational issue for them), many industry representatives—especially associations representing firms engaged in the export of scrap metals and other materials for recycling—continue to lobby hard against the imposition of the recycling ban. In addition, some countries—India, for example—are threatening to break ranks on the recycling ban.[47]

Regulation of the Hazardous Waste Trade within the EC/EU
The period during which the waste trade began to emerge onto the international scene also witnessed the phase of European integration that culminated in the Single European Act of 1987 (implemented 1992)

and the 1991 Treaty on European Union, also known as the Maastricht Treaty. At work in Europe at this time were two dynamics whose contradictory nature has provided the major theme for the literature on the environmental impacts of European integration.[48] First, the lowering of barriers to the movement of goods, capital, and labor also facilitated the movement of pollutants across national borders, through the twin effects of economic expansion and lower controls over pollutants transported across borders via human agency—for example, hazardous chemicals and wastes. Second, and partly in response to this, the EC began to take seriously the need to foster cooperation among its member states in a region highly vulnerable to transboundary environmental degradation, shifting from its earlier position that national environmental regulations constituted nontariff barriers to the free movement of factors of production.

The first dynamic—the rise of free trade in the EC—had an important effect on the waste trade between West European states because it facilitated the transfer of wastes across national frontiers (Laurence and Wynne 1989). At this point in time, wastes were still regarded as "normal goods" under trade legislation, and such EC-wide regulation that existed sought more to monitor, rather than restrict, the movement of wastes across national boundaries.

The more activist role taken by the EC in setting both the content and the processes of its members' environmental policies began in the 1970s, when DG XI began issuing environmental directives.[49] The interest of the EC/EU in hazardous waste management issues was first sparked by the 1976 Seveso Incident, when several drums of dioxin vanished from a plant in Italy following a severe chemical explosion, only to reappear eight years later in a disused abattoir in France (Gourlay 1992:79). The discovery of these drums prompted an investigation into the implementation of the 1978 Waste Directive, which in turn led to much more attention being paid to Directive implementation in the EC as a whole (Haigh and Lanigan 1995:20).

In 1984 the European Community adopted the Directive on the Supervision and Control Within the European Community of the Transfrontier Shipments of Hazardous Wastes, which established a notification and authorization system for the transportation of hazardous wastes within the European Community.[50] A 1986 amendment addressed the issue of transport of hazardous wastes through a third EC country en route to its final destination (Handley 1989). As far as wastes are concerned, many fear that an open Europe will mean that wastes will travel unmonitored not only to traditional destinations but also to countries such as Spain and Portugal, which still lag behind the rest of Europe in the development of environmental regulations (Laurence and Wynne 1989).

Debates at the European level revolve around achieving national harmonization of waste definitions, whether or not hazardous wastes should be classified as "normal goods" for trade purposes, and whether the trade should be governed by the *self-sufficiency principle* or the *proximity principle*.[51] The European Commission leans toward the application (or attempted application) of the former—national self-sufficiency in waste management (Jupille 1996)—but several member states, including Germany, strongly support the latter, whereby wastes would be shipped to the nearest best disposal facility, regardless of jurisdiction. A 1991 amendment to the 1975 Directive on waste "now obliges Member States to establish an integrated and adequate network of disposal facilities. This network 'must enable the Community as a whole to become self-sufficient in waste disposal, and the Member States to move towards that aim individually, taking into account geographical circumstances or the need for specialized installations for certain types of wastes'" (ENDS Report 210, July 1992:38).[51] The EU also requires that all hazardous waste shipments must be covered by financial guarantees, although it is unclear who is actually responsible for these. Suffice to say, however, that EU policy addressing the control of transfrontier movements remains hampered by the failure so far to reach complete agreement on a harmonized definition of hazardous wastes, as well as by recent political controversies over the dismantling by Shell of the Brent Spar oil rig, and the conflicts between EU member states over the Packaging Waste Directive.[52] However, the EU also recognizes the importance of harmonizing national waste management practices, and to that end has established directives dealing with, for example, landfill and incineration techniques. In the final chapter of this book, I examine in more depth the evolution of EU environmental policy as it relates to changing national environmental policies.

Governing the Waste Trade: Implications of Problem Framing

Attempts by both the EU and the UNEP to regulate the waste trade suffer from certain problems related to their respective conceptualizations of the key problems associated with the trade. Controlling the export of wastes to less developed countries is a very important issue, but the waste trade regime is quite possibly destined to ultimate failure if it does not address the basic cause of such pratices. For most international policymakers, the waste trade is typically characterized as a "transboundary" (or trade) issue, in that the locus of the "problem" is the actual movement of the wastes across borders: the Basel Convention is one of the few international environmental agreements that does not set targets to reduce the pollutant being regulated. However, the above analysis suggests there is a strong case to be made that the waste trade is better viewed in the context of domestic problems of waste management—that is, as a "local cumulative issue" with transboundary consequences—since it stems from a growing inability of most industrialized countries to adequately handle the wastes they generate.

Typologies of international environmental issues based on issue area characteristics are common in the international environmental politics literature (see chapter 3). One of the central propositions of the academic field of international environmental politics so far is that the form of international environmental agreements is driven by the structural characteristics of the issue in question (Downie 1994). Furthermore, the closer the fit between issue characteristics and institutional design, the more effective the regime is likely to be. For example, negotiations over ozone layer depletion involved multilateral bargaining among many countries, and set up a structure that not only included a phase-out timetable for the production of ozone-depleting substances, but also contained mechanisms for financial transfers to countries needing help in covering adjustment costs to safer products.

Following this division, there are two ways of framing the hazardous waste trade as an environmental issue. International agreements governing the trade focus almost wholly on the transportation of wastes across borders (particularly from developed to less developed countries), in terms of monitoring, facilitating, restricting, or banning the trade. It is possible, however, to conceptualize the waste trade very differently: as a function—indeed, the most visible symptom at the international level—of domestic-level problems of waste management shared, although in different forms, by all industrialized countries. This framing of the problem leads, for example, to analytical parallels with the trade in tropical timbers, and how that trade relates to tropical deforestation—parallels that are more apt than those with the "classic" transboundary issues of long-range air or river pollution.[53] Conceptualizing the waste trade as a transboundary issue addresses the most immediate aspects of the problem: monitoring and controlling wastes in transit and preventing waste dumping on poorer countries. However, a ban on waste trading, as exemplified by the new terms of the Basel Convention, fails to address the full significance of hazardous waste management problems embodied in the trade. Thus a strong case can be made in support of the latter way of framing the issues involved.[54]

The transboundary view of the waste trade treats waste disposal as a separate issue from waste production. By focusing solely on the downstream end of the waste life cycle, it ignores the possibility for enacting effective controls at the upstream end. Furthermore, waste management systems in developed countries are currently being pushed to capacity, and whereas the waste trade has acted as a safety valve in the past, current international controls are making that option (appropriately, in many cases) less available. More significantly, current trends predict that as the newly industrialized and less developed economies grow, they will increase their production of industrial and hazardous wastes. In the absence of effective environmental regulations, including regulations concerning waste management systems in these countries, the "waste problem" threatens to become truly global in scope.[56] To be successful, the international regulatory regime needs to focus on 1) waste minimization strategies across all countries, and 2) the development and strengthening of waste management systems, particularly in industrializing countries, ensuring that all countries develop and/or maintain sound waste management policies. In the following chapters I build on these insights to explain why some industrialized countries import hazardous wastes and others do not.

Conclusions and Next Steps

The vast bulk of the legal trade in hazardous wastes occurs among member states of the OECD. In most of these countries, the issue of waste management at the domestic level has become extremely contentious in recent years: not only are waste disposal facilities reaching full capacity, but public perceptions are that the risks associated with hazardous waste management are very high. In this context it is not hard to explain why some countries choose to export part of their wastes. The more difficult question is why some industrialized countries continue to accept waste imports—imports associated with a high degree of risk to human health and to local ecosystems—in the absence of coercion or the sort of financial incentives that have affected the decisions of many less developed countries.

The three possible answers to this question correspond to the contending explanations outlined in chapter 1: regulatory differences, financial incentives (or economic nationalism), and comparative advantage. First, no risk is being taken: wastes are exported to countries with superior disposal facilities—for example, specialized incinerators that can safely destroy highly toxic substances. This argument corresponds to the comparative advantage explanation. Second, the risk is there, but it is small and, furthermore, far outweighed by the potential benefits—for example, the financial rewards reaped. This argument corresponds to the financial incentives/economic-nationalist explanation. Third, imported wastes do in fact pose a considerable degree of risk to the general population and to the environment of the importing country, but although public opinion often runs strongly against waste importation, other factors enable firms within that country to engage in waste importation. I contend that this third answer is the correct one, and that these "other factors" consist of differences between national systems of environmental regulation, as I explain in chapter 3, and as chapters 4, 5, and 6—on the waste trade and management policies of Great Britain, Germany, France, Japan, and Australia—demonstrate.

The international regime governing the waste trade has evolved rapidly, and legal embargoes on all exports of hazardous wastes to less developed countries are strongly advocated by significant sectors of the international community. The arguments in this chapter suggest that in the absence of a concerted effort on the part of developed countries to change their waste management and generation practices, a ban on waste trading is likely to be at best ineffective, at worst disastrous. The waste trade is conceptualized here as a problem of the developed countries that may come to be shared by others as they industrialize further. I return to the policy implications of this view in the final chapter. In chapter 3 I situate this approach in the existing literatures on international and comparative environmental politics, and I begin outlining how the questions raised here can be formulated and answered.

3
An Institutional Explanation of the Waste Trade

Many domestic-level explanations of international political outcomes focus on the actions and preferences of different domestic-level actors—such as state actors, societal interest groups, or firms—and how these affect government policy.[1] In contrast, a growing body of literature argues that it is necessary to examine the institutional structures that pattern relationships between state and societal actors, the goals they pursue (and the means of pursuit), and the power relations between them. As they persist over time, these structures (or social "ground rules") constrain—and indeed even change—the range of options available to actors within the system, hence influencing policy outcomes.[2]

The arguments I present here support this institutional approach and extend it to the arena of international environmental politics. They focus on a particular set of rules concerning environmental regulation: so-called intermediate-level institutional factors (Thelen and Steinmo 1992:6). Specific features of national systems of environmental regulation profoundly affect a country's propensity to legally import hazardous wastes from abroad, and differences between these systems help determine patterns of trade in hazardous wastes among developed countries. In this chapter I develop this book's theoretical framework; I also discuss relevant hypotheses and the methodological issues involved in selecting cases, finding data, and determining the period of time over which this study extends.

Causal Links between Regulatory Systems and the Waste Trade

Regulation, defined as government attempts "to control some private sector economic decisions to which the government is not a party" (Noll 1985:9), takes many forms, dependent in part on the practice being governed and in part on the political system of the country concerned.[3] The basic functions of national environmental regulations are to control and monitor practices that lead to environmental degradation, to reduce degradation at the national level, and to repair past environmental damage. The theoretical importance of environmental regulations at the domestic level lies in the way that these institutional rules shape the relationship between public and private sector actors in a given country: what might be termed a country's regulatory climate. Some regulatory systems allow private actors who might be engaged in polluting activity—such as firms—greater leeway in their actions, in terms of importing higher or lower levels of waste; other systems allow much less freedom.

Thus the argument presented here concerns the ways in which preferences of different actors who have a stake in furthering or controlling the waste trade—government officials, waste disposal companies, and other societal actors (environmental groups and the greater public)—are translated into policy outputs. This process of transmission is shaped fundamentally by the institutional mechanisms that make up a country's regulatory system. These act to filter out or to enhance the influence of different societal groups at different points in the policy process. They also fundamentally affect the degree to which government actors in the system can monitor and/or control the actions of the firms and individuals under their jurisdiction once policy goals have been determined.

A country's propensity to import hazardous wastes is not necessarily related to the effectiveness or success of its environmental regulation. Rather, it says more about its willingness to take on environmental risks that others are not willing to shoulder. In an increasingly open international economic system, where states are extensively linked through trade and other economic agreements and national borders have become more permeable, differences between regulatory systems have become much more important in determining the volume and direction of flows of goods and factors of production. Despite the ways in which the waste trade differs from trade in "normal" goods, a similar analysis applies: while some regulatory climates are more conducive to the exportation of wastes, others are likely to lead to importation. These differences interact at the international level to determine specific directions of waste movements—for example, Great Britain is a net importer of wastes and Germany is a net exporter.

Dependent Variable: National Propensities to Import Hazardous Wastes

The dependent variable for this study is a country's propensity to legally import hazardous wastes, measured in terms of volumes of wastes traded and their annual net imports of wastes over a number of years.[4] Legal waste shipments are those officially approved by importing authorities in accordance either with domestic or international regulations. As noted in chapter 2, roughly 80 percent of the international waste trade—trade that occurs between OECD countries—is legal by this definition, yet it has received much less attention than illegal trade in the relevant literature (see Montgomery 1995).[5]

The legal hazardous waste trade is a good case for this sort of analysis for a number of reasons. First, it is a highly salient and politicized issue in the international arena. Second, by nature it is agency driven: someone makes a decision to export or import wastes. Thus it is (theoretically) possible to find out who makes these decisions, and at what level. Third, close examination of countries' involvement in the trade brings up a number of empirical puzzles worthy of attention, in particular, why some advanced countries, such as Britain or France, import wastes despite objections from their populations, while others, such as Germany or Australia, are net exporters. In each of these cases, it is possible to identify similar configurations of actor preferences, yet the markedly different outcomes suggest the operation of other factors. Furthermore, national policy responses differ as well: all of these countries have announced a desire to ban the waste trade, and some have even instituted policies to this effect. However, some governments are proving more effective than others in carrying through these intentions.

More generally, waste importation imposes a reasonably high degree of (actual and/or perceived) environmental risk on recipient populations and local ecosystems. Hence it provides an informative and measurable proxy variable in identifying and comparing countries' risk attitudes toward international environmental degradation. Finally, examining a country's involvement in the waste trade provides insight into the operation of its regulatory system, perhaps more so than simply examining a country's level of success in dealing with pollution. One reason for this, in addition to providing a relatively easy to measure dependent variable, is that the waste trade has in most cases been an unintended spinoff of other policies; only very recently have developed countries begun to formulate policies specifically addressing the trade.

Institutional Explanation: A Nested Model of Environmental Regulation

Hazardous waste regulation began in most industrialized countries in the early-1970s, when specific practices and rules for dealing with such substances were put in place for the first time.[6] However, these practices cannot be seen as a simple technocratic response to the specific problems of dealing with hazardous wastes. Rather, they are embedded in and cannot be separated from broader national practices of environmental regulation driving policy formulation and implementation across environmental issue areas.[7] Weale (1992a) identifies several distinctive characteristics of pollution that have driven policy formation in this area.[8] In turn, he argues that it is not possible to understand how national practices have evolved and how they operate without examining the broader political context in which state and societal actors interact, and how this interaction has shaped certain distinctive national responses to environmental degradation. Wynne emphasizes the "interaction between technical and institutional factors in regulation": a system of regulation cannot be understood simply by reference to its technical specifications, but must be seen in the context of the social and political rules and norms that underlie it (Wynne 1987:405). The theoretical approach I develop here therefore integrates these nested levels of hazardous waste management and environmental regulation: rules governing hazardous waste management, systems of environmental regulation, and wider political institutions and processes.

The complexity of the sets of rules and practices that make up most systems of environmental regulation make it impossible to generalize about their effects without breaking them down into their component parts. Simply stating that one country's environmental regulations are more lax than another's will not suffice to explain the complex interactions and effects making up an environmental regime, be it national or international.

Certain characteristics are shared by most systems of environmental regulation. These include the goals of regulation, the actors being regulated (in the hazardous waste case, firms that generate the waste and waste disposal companies), and the sorts of instruments (or policy mechanisms) applied, such as emissions standards and controls, fines, and inspections. All the countries under investigation have advanced systems of environmental regulation in place.[9] Thus the significant differences lie not in the substantive content of these regulations, nor even in their effectiveness, but rather in the ways these policies are formulated, enacted and implemented, and followed by the actors being regulated.

To this end, the most important factors in understanding waste importation practices are a country's regulatory *structure*—the allocation of responsibilities among different agencies or levels of government—and its regulatory *style*—how policy is made and implemented. These variables work together in two ways. First, they determine the degree of coordination in the system and the extent to which waste importation practices can be monitored and controlled by different levels of regulatory authorities: a principal-agent argument.[10] Second, institutional structures and regulatory styles filter the preferences of different groups of stakeholders in the waste trade at different stages of the policy process, thus determining policy outcomes with respect to waste importation.

Structure of Hazardous Waste Management and Regulation

Responsibilities for the management and regulation of hazardous wastes may be allocated across several different levels or agencies within a national system. Regulatory structure, as defined here, consists of three dimensions:

1. the structure of the waste management industry—in particular, whether it is privately or publicly owned—and its degree of competitiveness,
2. the allocation of regulatory responsibilities among different levels of government and among government agencies at the national level,
3. whether the overarching polity is federal or unitary.

How a state falls along each of these three dimensions defines the degree to which a regulatory structure is centralized or decentralized, taking into account the number of actors in the system and the degree of coordination among them. Regulatory structures may be classed on a continuum between fully centralized and diffuse. A diffuse system is one where regulatory responsibilities are divided among many different agencies at both the national and local levels, and where the waste disposal industry is privately owned and highly competitive. A fully centralized system is, in contrast, one where a single government agency holds all waste management responsibilities, and where the waste disposal industry is a public monopoly.

Regulatory structure matters in that it affects the ability of government officials to monitor, control, and coordinate the actions of its citizens: the less coordination there is in the system, the more leeway firms have to import hazardous wastes. Following principal-agent theory, the more decentralized a system is, the less likely are the principals (national government officials) to be able to monitor and affect the behavior of the agents (for example, local authorities' issuing of importation permits and/or firms' importing of hazardous wastes). Furthermore, the more actors there are within the regulatory structure and the more levels that exist in the chain of command, the more complex the monitoring problem becomes. This problem in turn may be exacerbated by the degree of uncertainty in the system regarding the respective roles and responsibilities of the various actors.

A fully centralized system, including public management of waste disposal facilities, is usually considered to be the most effective in controlling waste importation and, more generally, in implementing policy goals. The perceived advantages of a centralized system are that it leads to lower transactions and information costs and that there is a core group of officials directly responsible for the protection of the environment.[11] This generates the following hypothesis, whose underlying logic I discuss in the following section:

H1: *The more decentralized a country's structure of hazardous waste management and regulation, the more likely it is to import hazardous wastes.*

Structure of the Waste Management Industry Linnerooth and Davis (1987:165) identify four key variables in assessing the structure of a country's waste disposal industry:

1. the extent of public versus private ownership and control,
2. a monopolized versus competitive market,
3. integrated, comprehensive facilities versus segregated, specialized facilities or onsite management,
4. cost allocation between industry and the public purse.

The first two provide the basis for distinguishing three basic models for the management of hazardous wastes (ibid.:169). The first is the "public monopoly" model, exemplified by the Bavarian and Hessian waste management systems, where the waste disposal industry is concentrated into a single plant under federal or state government control. At the other extreme is the "private competitive" model, exemplified by the United States, where a large number of private waste disposal firms compete for contracts. In between lies the "public utility" model, which contains different combinations of private and regulated monopolies. Variations within these basic models depend on the latter two variables: the degree of facility integration and the extent of government subsidization or taxation.

Ownership of the waste management industry is itself a function of government policy and a country's regulatory culture, a reflection of beliefs about the most effective means of provision of essential environmental services (Linnerooth and Davis 1987; Majone 1996). Many scholars and policymakers ascribe to the view that waste regulation is essential to alleviate problems of negative externalities and public goods provision associated with waste disposal. The separation of waste disposal and waste regulation responsibilities generates a whole set of principal-agent relations, with attendant opportunities for the agents—firms—and dilemmas for the principal(s), who are likely to have a different set of preferences than firms and to be hampered by lack of access to firms' proprietary information.

The more competitive a waste disposal industry, the more likely it is to become attractive to waste generators, both at home and abroad.[12] It is also likely that waste disposal companies operating in a competitive environment will make a deliberate effort to attract waste imports from abroad or to become fully multinational. This issue is compounded by that of facility ownership: privately owned firms are likely to place a higher value on maximum profits than general goals of social welfare; publicly owned firms usually operate under a different mandate.

Allocation of Responsibilities across Agencies and Levels of Government
The responsibilities for planning, formulating, and implementing a country's environmental regulations may be centrally coordinated or divided across different regulatory bodies (Jordan 1993:408). In some countries, regulatory responsibilities lie wholly with a single national agency or ministry for the environment, and the "purest" type of centralized system is one where pollution control policy is fully (internally) integrated.[13] In most countries, however, regulatory responsibilities are allocated across different levels of government and/or across different agencies. Although most industrialized countries have an environment ministry, often other ministries (agriculture, industry, or health) also have a voice in the making of environmental policies.[14]

When regulatory responsibilities are divided between national and local government (as is nearly always the case), federal and unitary (nonfederal) systems behave very differently. Under federal constitutions, subunits have considerable constitutionally protected powers, responsibilities, and sources of revenue. A unitary system is one where ultimate political authority belongs to a central government. Under such a system, however, issues that are seen to have local effects are commonly delegated to lower levels of government, such as local authorities in the British case. Conventional wisdom holds that federal systems are typically less effective than unitary systems at implementing policy, owing to the more fragmented nature of the political process.[15]

In this study I advocate the more counterintuitive proposition that federal systems are better able to manage hazardous wastes and waste trading than unitary systems, where environmental protection is primarily in the hands of local authorities. First, conventional wisdom tends to ignore the fact that environmental regulatory responsibilities are rarely vested in a single, central political authority, but are instead devolved to multiple local authorities. In the British case, for instance, hazardous wastes have traditionally been considered a local issue in terms of their effects and hence best dealt with at that level. In fact in several of the cases in this book, responsibility for issuing waste importation permits lies at the local rather than the national level. Second, there are important political differences between the two sorts of systems that make it much harder for authorities in a unitary system to coordinate and control activities such as waste trading. Local authorities in unitary systems do not have the same autonomy or protection of their role as do subunits in federal systems, and center-local relations are often more uncertain. In addition, local regulatory authorities are often dependent on the whims of central government, and as a result must compete for scarce funding or seek other means of raising revenue.

When central government is responsible for the control of local regulatory authorities, several factors come into play. First, if the "rules of the game" between central and local governments are unclear, and local authorities are financially insecure or even worried about their continued existence, it is unlikely that central government will be able to fully control the activities of local regulatory bodies. Information asymmetries will exist on each side, fueled by a prevailing climate of mistrust, where formal mechanisms for exchanging information are absent. For example, local authorities in this situation have no incentive to reveal their preferences or activities to central government. Conditions where central and local preferences conflict are arguably much more likely to exist in unitary than federal polities. Similarly, when local authorities that depend on waste disposal as an important source of revenue must compete for scarce financial resources, they are more likely to turn to waste importation as an additional means of raising revenue. For example, states and local authorities seeking to build up their industrial base may seek to attract both waste disposal and waste generating industries (Bowman 1985:109). Again, this situation is arguably more likely in nonfederal systems, where local governments and authorities tend to have less ability to raise their own revenue and less claim to a broad base of legitimacy from society to combat funding cutbacks from central government.[16] Finally, where control over the movement and importation of wastes is relegated to local authorities, officials are unlikely to have the capability to find out whether the movement of waste is warranted under international law. They also lack the jurisdictional capability to discriminate between wastes imported into their territory, be they from a bordering area or from the other side of the planet. This again gives firms more incentive to import wastes from abroad.[17]

Indicators of Regulatory Style

The concept of regulatory style has a long academic pedigree. In a study of cross-sectoral national styles, Richardson defines it as "the interaction between the government's approach to problem solving, and the relationship between government and other actors in the policy process" (Richardson 1982:13) or "'standard operating procedures' for making and implementing policies" (2). In one of the leading studies in this field, David Vogel defines environmental regulatory style as the strategies governments adopt "for improving the quality of their physical environment and safeguarding the health of their population" (Vogel 1986:20–21). Hence the concept covers the process of policy formulation and implementation, but does not include a measure of policy success.

There are several different ways to categorize countries' styles of environmental regulation. For example, Vogel compares the more rigid and rule-oriented and more publicly conflictual style in the United States with the more flexible and informal style in Great Britain (Vogel 1986).[18] Susan Aguilar, discussing Germany and Spain, distinguishes between corporatist and conflictual styles of environmental regulation.

The corporatist style of regulation is characterized by a leading role taken by industry in policy negotiations and by a reliance on voluntary agreements, while the conflictual style is characterized by the leading role taken by government or state actors to the exclusion of other interested parties (Aguilar 1993). Van Waarden (1995), in his review of the relevant literature, identifies no fewer than six contrasting types of policy style.[19]

If one accepts that identifying a country's regulatory style involves understanding how government actors interact with different societal groups in making and implementing policy, then it becomes necessary to break the concept down into its constituent parts. Government-industry-society relations in the policymaking process are conditioned by the various points of access accorded to different groups. Policy implementation is to a greater degree determined by the "standard operating procedures" followed by government actors. These are the two factors most important in identifying a country's regulatory style, and I discuss them in depth in the following sections.

Determinants of Access to the Policy Process The first element—*policy access*—revolves around the configuration of state-business-society relations within a given country.[20] Each of these groups has certain interests in environmental policy, and these interests often conflict. How these conflicts are resolved depends crucially on which group has the greater access to and influence on the policy process. This in turn depends, at least in part, on the broader institutional framework— which channels or filters demands from different groups—of the country involved.[21]

Three broad groups of actors comprise the main stakeholders in the waste trade, in terms of either wanting to continue or to prevent the trade. State (or government) actors are those officials and elected representatives responsible for making and implementing environmental regulations in general and hazardous waste policy in particular. The most important business interest is the waste disposal industry, which may or may not act in conjunction with other industries in trying to influence environmental policy. Where waste exportation dominates importation, the interests of firms that export wastes (including the waste generators) also have to be taken into account. "Society"—outside of the waste disposal industry—is a harder group of actors to define or single out. In a necessary simplification, this study identifies first, public opinion in general, and second, environmental pressure groups and political parties as stakeholders in the trade.[22] There are established literatures on environmental groups and public opinion with respect to the environment across most industrialized countries.[23]

Countries can be placed along a continuum of styles with respect to policy access: at one end of the continuum are *open* (or *inclusive*) styles, and at the other, *closed* (or *exclusive*) styles. Open styles of regulation are those where all groups have access to the policy process at its different stages. This might be through the electoral process, the court system, or widespread public consultation during the policy formation stage. Fully closed styles are those where the state acts autonomously in the formulation and implementation of environmental policy. An intermediate case is where the state works in close collaboration with business to the exclusion of other constituencies.[24] In the closed and intermediate cases, the policy process tends to be secretive, and there are few institutional mechanisms for enabling widespread public participation in the process. Therefore countries with closed styles of regulation or those that favor the interests of business are more likely to import hazardous wastes than those characterized by a higher degree of public access, as public opinion is likely to run strongly against waste importation while business interests are likely to favor it.[25] Hence:

H2: *The more closed a country's system of environmental regulation, in terms of allowing access to the policy process to a wide range of groups and interests, the more likely it is to import hazardous wastes.*

Understanding how and where access to the policy process is achieved means examining both the policy communities in which deliberation occurs and countries' broader political opportunity structures. Combining these two allows an assessment of how many points of access exist within a system and to whom those points of access are accorded. Policy communities are "highly integrated policy networks" around specific policy sectors, comprising "ministers, key civil servants ... and the leaders and officers of key interest groups" (Winter 1996:25–26) with strong shared interests and centripetal tendencies.[26] Obviously important for this analysis is identifying who the "key interest groups" are (for example, scientific interests, business groups, or environmental organizations) and how their representation in the policy process is determined (e.g., at the whim of the governing party or via formal rules or procedures).

According to Kitschelt, "political opportunity structures are comprised of specific configurations of resources, institutional arrangements and historical precedents for social mobilization, which facilitate the development of protest movements in some instances and constrain them in others" (Kitschelt 1986:58).[27] This notion, slightly modified, can be usefully employed here to capture the effects on the policy process of the broader political, economic, and cultural context in which policy responses are determined, and nowhere is this more true than in the issue of regulatory style. This study identifies three basic constitutional parameters of each case as key indicators of a country's political opportunity structure: how interests are represented in the policy process, the role of the judiciary, and whether the state is federal or unitary.

Mode of Policy Implementation The second element concerns a country's prevailing mode of policy implementation, which may be rigid or flexible.[28] *Rigid* systems are characterized by strict, nationally applied standards of pollution control—for example, a Clean Air Act that strictly limits the quantities of emissions allowed into the atmosphere. These standards are applied without discrimination across industries and are often judicially enforced (fines are a common punishment mechanism). *Flexible* systems, on the other hand, apply regulations on a case by case basis. Rather than relying on nationally applied technical standards, they tend to work on the principle that industries and firms are individually responsible for choosing the best means possible for controlling their effects on the physical environment. Flexible systems rely on a fairly high degree of self-enforcement by industry, and legislation might be tailored to allow for "loopholes" that may be taken advantage of by firms—a practice many defend by claiming it reduces the incentives for firms to evade or circumvent the legislation entirely. In addition, flexible systems typically have much smaller bureaucracies than rigid systems, in terms of budget and personnel, to enforce legislation.

Table 3.1
Summary of hypotheses related to environmental regulation

Structure	Access	Implementation	Import propensity
Diffuse	Closed	Flexible	High
Centralized	Open	Rigid	Low

Flexible styles of regulation create a climate much more conducive to waste importation. Firms are under less strict requirements regarding pollution and the acceptable risks they can take with respect to the environment, and such systems tend to lack the central coordinating mechanisms and rules necessary to fully monitor and control the movement of hazardous wastes. This yields the following hypothesis:

H3: *The more flexible a country's mode of policy implementation, the more likely it is to import hazardous wastes.*

Table 3.1 summarizes the ways in which countries vary in regulatory characteristics. As it shows, waste importation propensities can only be definitively *predicted* in the two extreme cases. However, by coding each country in the empirical analysis according to how it fits in each category, it should be possible to *explain* their importation propensities. Except to a limited extent in chapter 6, I do not test the three hypotheses presented here against each other. Rather, in combination they make up a picture of procedural aspects of a country's regulatory system. There is also no a priori reason to expect them to covary—for example, to expect that countries with centralized structures also employ rigid modes of implementation.

Alternative Explanations

To demonstrate the explanatory power of this argument, two tasks must be accomplished in the empirical analysis. The first is to validate the institutional argument on its own terms, that is, to show that regulatory structures have a determining effect on the choices of actors involved in or affected by the international waste trade, rather than vice versa. The second task is to evaluate the arguments concerning environmental regulation against alternative explanations. Three types of alternative explanations are identified here. International approaches, as explained below, do not work well in this context. However, two alternative explanations deserve further attention. One is based on the preferences of state actors: an economic nationalist approach. The other is based on whether a country has a comparative advantage in the provision of waste disposal facilities.

International- versus Domestic-Level Approaches
There are several reasons why international-level explanations do not explain specific patterns of waste trading among developed countries. An argument could be made that the waste trade is a function of the dense set of transnational linkages—over and above their membership in the (self-described) "club" of industrialized nations—within which the OECD nations interact.[29] These include regional and international trade agreements, military or security alliances, former colonial ties (the Commonwealth, for example), and membership in the European Union. Such linkages facilitate the waste trade by creating a set of relationships that spill over into other issue areas, leading to a situation where waste disposal contracts are more easily negotiated (Laurence and Wynne 1989). Hence the countries involved are more likely to be engaged in waste trade with each other than with countries outside the group, and, indeed, such processes are important in understanding why intra-OECD waste trading arose in the first place. However, given the directional focus of the dependent variable, this factor cannot explain the actions of specific OECD countries—why, for example, Great Britain imports wastes, while Germany, which shares the same set of linkages, exports them.

Likewise, an explanation based on the terms of the international regime governing the trade is of little help. First, the main provisions of the Basel Convention are still coming into effect. More fundamentally, this regime is concerned not with the intra-OECD trade, but with the trade between OECD and non-OECD nations.[30] Finally, realist theorists might argue that Britain imports Germany's waste because Germany is stronger than Britain and therefore able to dump its wastes on its weaker partner. Germany is, after all, considered the most powerful nation in Western Europe, and certainly is better off than Britain in terms of GNP. However, the differences are not as extreme as, say, those between Germany and its eastern neighbors. Such an approach also cannot capture the full dynamics of waste trading—for example, the role played by firms. Hence it is necessary to turn to the differences that exist between states at the domestic level in order to explain different levels of engagement in the waste trade.

State Calculations of Costs and Benefits Associated with the Trade
The first contending explanation is based on Montgomery's economic nationalist explanation (Montgomery 1992, 1995), which focuses on the way national governments assess the costs and benefits of hazardous waste importation. First, the waste trade generates a lot of money. These revenues accrue on the whole to the companies disposing of the wastes. However, there are two main ways in which governments in Western Europe, in addition to the importing firms, may benefit from allowing these transactions: through a contribution to a country's balance of payments, its employment levels, and the possible imposition of tariffs or taxes on imported wastes. Balanced against these calculations of potential benefits are the risks associated with waste importation, including environmental risk and the level of domestic opposition. This argument contends, therefore, that waste importation schemes are approved on an individual basis, based on the calculation by state officials of the potential costs and benefits of each transaction. How these fall determines the extent to which a country engages in waste importation.

The power of this explanation depends on three factors. First, it must be demonstrated that state actors act in a unitary (and rational) way when making these calculations—in other words, that they are not subject to external constraints or influences in making decisions regarding the waste trade. Second, it must be shown that the associated risks and benefits can actually be calculated in a satisfactory way, and that the political costs of allowing, say, waste importation, do not outweigh the economic benefits. Third, this argument has to stand up to comparative analysis, explaining why, if the waste trade is so lucrative, some countries choose not to import wastes—especially, as is the case for OECD countries, if their economic positions are relatively similar.

Comparative Advantage in Waste Disposal

It could be argued that some countries have a comparative advantage over others in disposing of wastes: either they can dispose of wastes more cheaply than other countries or they have the facilities to deal with certain types of waste.[31] Most OECD countries have an extensive network of commercial or specialized waste disposal facilities. Two factors in particular explain why a country's waste disposal industry may create a climate conducive to waste importation: total disposal capacity and the level of disposal technology.

Disposal Capacity Presumably, countries with higher excess waste disposal capacity once domestic needs are met are more likely to import hazardous wastes. Total waste disposal capacity is measured in terms of the number of tonnes of waste a country can dispose of on an annual basis, excluding onsite disposal facilities, as these are unlikely to be used for the processing of imported wastes.[32]

Accurately assessing this variable is difficult, as most available data put out by OECD countries reports only the amount of wastes actually disposed of, not the total available capacity. There are two possible proxy measures where this data is not available: the number of disposal facilities a country possesses and general trends in total waste disposal capacity, information on which can be obtained from some sources. For example, total capacity has declined in a large number of industrialized countries as a result of the decreased use of landfill as a disposal technique (Hilz and Ehrenfeld 1991). In such cases, this variable can be controlled for by assuming that most OECD countries operate at or close to full capacity.

Disposal Technology While any country with a patch of waste ground has a potential hazardous waste disposal facility, there are many different techniques for the disposal of hazardous wastes, varying extensively in terms of level of technology, cost, and effectiveness in disposing of harmful elements. Most analysts list disposal techniques in order of effectiveness in disposing of harmful elements—the waste management hierarchy discussed in chapter 2.

It is hardly surprising then that propensity to import hazardous wastes is profoundly influenced by the type, and quality, of disposal facilities available. There are two schools of thought on this issue. The first argues that wastes travel to countries with more advanced systems of disposal. This argument maintains that waste disposal is a service industry, and that the main concern of waste producers is that wastes are disposed of in an environmentally sound manner. Certain countries are therefore able to build up a comparative advantage in waste disposal based on the quality of services they offer. Recycling and incineration or chemical treatment facilities are considered most effective in treating hazardous substances, landfill sites the least.[33] Level and quality of disposal technology is the basis of the regulatory principle underlying the Basel Convention, which states that waste should not be exported to countries with facilities inferior to those of the exporter.

The second school of thought argues that wastes are much more likely to follow the path of least cost, in strictly short-run financial terms: producers concerned by the costs of waste disposal at home are likely to search for cheaper options abroad. If the unit cost of waste disposal is positively correlated with the quality of technology, this argument is the converse of the first: potential importers will seek to build up a comparative advantage in waste disposal based on the relative costs of waste disposal.[34] The technologies surveyed here vary widely in terms of both fixed and variable disposal costs. Incineration facilities are the most expensive, both to set up and to operate, while landfill sites, especially uncontrolled sites, are the least expensive.[35]

The question now is which version should be used as the basis for this analysis. With respect to legal trade among advanced industrialized democracies, the first argument—that wastes should travel to countries with more advanced disposal systems—is more compelling for two reasons. First, government officials when deciding to import wastes should take into account long-term costs—for example, the costs of cleaning up waste dumps—before issuing import permits.[36] Second, countries that have agreed to abide by the principles of various international agreements have made public commitments to make their policy decisions according to those principles. Hence the argument presented here is that countries with more advanced disposal facilities are more likely to be involved in the legal importation of hazardous wastes.[37]

Methodological Issues

Case Selection and Hypothesis Testing

Testing the power of the domestic institutional argument against the empirical data involves two main steps. First, it is necessary to demonstrate in some depth how regulatory institutions constrain actors' behavior over time. This I accomplish by identifying the preferences and interests of actors relevant to the trade—government officials, the waste disposal industry, and other societal groups—and showing in detail how these are transmitted (or not) by the regulatory system into policy and waste trade outcomes: importation or no importation. The second step involves testing this argument against the alternative explanations identified above. In the case chapters I test the economic nationalist explanation on the British case, and the comparative advantage explanation on Britain and Germany.

The cases presented are the OECD countries Britain, Germany, France, Australia, and Japan.[38] There are several reasons for choosing these particular cases. First, they all have relatively sophisticated systems of environmental regulation—though these systems vary widely in terms of regulatory style, structure, and effectiveness—and prioritize environmental quality as a policy goal. Second, choosing this set of countries controls for a number of factors at the international level. These include levels of industrialization and involvement in international agreements affecting the waste trade. With respect to the hazardous waste trade, they are the countries most involved in the legal trade, not only as exporters of waste but also as importers. Finally, they are the countries on which most data is available, not only on their engagement in the waste trade, but also on their waste management systems, disposal infrastructure, and regulatory systems.

Chapters 4 and 5 deal with the cases of Britain and Germany respectively. This choice of cases controls for several potentially significant variables. Germany shares important characteristics with Great Britain, including membership in the European Union and status as an advanced industrialized democracy. They also share common responsibilities under the current international regulatory regime governing the waste trade. At the same time, the two countries display significant differences in terms of their domestic regulatory systems.[39] Britain has a diffuse regulatory structure, a closed style, and a very flexible mode of policy implementation. In contrast, Germany has a more, but not fully, centralized structure in the context of a federal polity, an open style of regulation, and a rigid mode of policy implementation.

In chapter 6, I extend the analysis to three other OECD countries—France, Australia, and Japan—in order to test the general applicability of the theoretical framework, and, to an extent, I test the three regulatory hypotheses presented in this chapter against each other. These cases were chosen primarily on the basis of observed variations in their regulatory systems.[40] France resembles Britain in many respects—for example, they are both unitary states with closed styles of regulation. It is also the world's largest importer of hazardous wastes. Australia on the other hand is, like Germany, a federal nonimporter of wastes with an open policy style. In addition, it is the only country in this study with a fully centralized waste management structure. Japan was chosen because it presents something of an anomaly for this argument: while its regulatory structure and style matches Britain's to the extent that one would expect it to import wastes, it in fact does not. I examine why this is the case. The United States was not chosen as a case, in part because its engagement in the waste trade differs fundamentally from other OECD countries and in part because its hazardous waste regulations differ considerably from state to state.[41]

Periodicity
This book chiefly addresses the period from the 1980s up to the middle- and late-1990s. Over this time, the countries I examined maintained relatively stable waste trade and regulatory practices. This is not to say that dynamic elements are completely absent: most of the countries under examination have exhibited some form of regulatory change, or dynamism, over this time period as a result of either internal or external pressures (Dyson 1992; Thelen and Steinmo 1992). Important internal sources of change are often related to changes in government, particularly when they also involve substantial policy transformation,

demands at the societal level for environmental protection, or significant advances in regulation techniques. External pressures include crises or disasters, economic forces for regulatory harmonization, and the impacts of international organizations and agreements. The most important international actor in this respect has been the European Union, which has sought to change not only the targets of its members' regulations but also the way in which they conduct the policy process.[42]

However, as Jordan (1993) argues, countries are generally slow to adapt to pressures for change, changes that do occur tend to be incremental rather than radical, and countries vary extensively in their degree of resistance to external change. Even where radical change has been proposed—as with Britain's attempts to move toward a system of integrated pollution control—progress has tended to be slow and piecemeal. While the theoretical approach I employ here lends itself to diachronic analysis (changes over time), I have chosen to keep this study within the limits of synchronic analysis, examining the impact of specific regulatory arrangements on waste trade practices. In each chapter, however, I do examine key trends in environmental legislation overall and in hazardous waste management in each case. In the final chapter I examine arguments concerning pace and direction of regulatory change, and the future of the waste trade in the context of more detailed and prescriptive regulations at the international and European levels. I also look at how different approaches intersect with or are augmented by an institutional approach such as that taken here.

Data Issues

Collecting reliable and consistent data on the hazardous waste trade is not easy.[43] The biggest problem is that there is no internationally accepted definition of what constitutes a hazardous waste, or even a listing of toxic substances to be regulated.[44] Furthermore, the data that has been gathered and tabulated goes back only to 1988; earlier reports are usually anecdotal. Thus the figures reported by national governments are based on national definitions and, as the OECD correctly argues, to be treated with caution.

There are, however, several main sources of data on hazardous waste trading, generation, and disposal that do allow for generalization and comparison, and in combination enable the coding of countries into categories of either waste importer or exporter. First and foremost, the OECD and the different governments produce data sets on a more or less annual basis. The OECD has also produced a series of environmental performance reviews that provide overviews of the environmental regulatory systems of the member states.[45] Second, there are several trade periodicals on aspects of hazardous waste management and environmental policies: the most useful of these are *Haznews*, the *ENDS Report*, *Waste Management and Research* (produced by the ISWA), and, on Germany, *Müll und Abfall*. For historical analysis, the two most important studies are Forester and Skinner 1987 and Wynne 1987. Finally, Greenpeace has actively campaigned against the waste trade, and its most recent reports (Vallette and Spalding 1990; Heller 1994)—although they focus mainly on waste exports from OECD to non-OECD countries—provide useful comparative figures to support or, in some cases, highlight discrepancies with those put out by the OECD and national governments.

Where possible, it is useful to identify whether a country imports wastes primarily for final disposal or for energy/materials recycling and recovery. However, the significance of this variation is hard to detect for a number of reasons. Many countries do not report this distinction, reporting only imports of wastes destined for final disposal. In addition, the loophole in the international regime that has allowed transfrontier movements of wastes for recovery has provided waste brokers with a variety of creative labeling practices that do not necessarily reflect the intended ultimate destination of the wastes.

Conclusion

The role of institutions in constraining or facilitating certain political outcomes above others is coming to be a common theme in comparative politics and international relations. Determining the effects of "nontangibles" such as regulatory style and structure is not easy, requiring as it does careful analysis of how different preferences and interests are filtered through the policy process. The next few chapters tell the story of waste trading in Britain, Germany, France, Australia and, Japan, focusing on the three explanations outlined in this chapter—institutional, economic-nationalist, and comparative advantage—and demonstrating the pathways through which institutional rules and practices affect whether countries import hazardous wastes. These stories combine to give a broader picture of regulatory politics and practices in this set of countries during this period—over and above the area of hazardous waste management—and of the constraints on the exercise of agency in democratic states. Above all, the following chapters show that hazardous waste management has not lost any of the controversial elements that gave rise to waste exportation in the first place.

4
Great Britain: Risk Acceptance and the Politics of Flexibility, Diffusion, and Closure

Why is Great Britain one of the world's largest importers of hazardous wastes? It has held this position throughout the 1980s and into the 1990s, during which time it has steadily increased the amount of wastes it imports from abroad. Nearly all of this trade is legal—conducted with the knowledge and consent of the British government—and the majority of these wastes come from other developed countries, especially Britain's partners in the European Union.

There are several reasons why this level of waste importation by Great Britain presents a puzzle.[1] First, Britain is a small and densely populated country, which would imply that its capacity for handling vast quantities of waste is strictly limited. Second, Britain is an advanced industrialized democracy with a sizable, although fragmented, environmental movement, and British public opinion has always been strongly pro-conservation. This implies, according to the "Not In My Backyard" phenomenon, that the strong public opinion opposing the importation of hazardous wastes should be translated into government policy, in terms of an advanced and efficient infrastructure for restricting their import. This translation has not yet happened. More interesting for scholars of international relations, Britain, in the absence of any form of coercion, economic or otherwise, deliberately takes on the risk of disposing of other countries' potential pollutants. This behavior challenges even the most liberal assumptions about state behavior in the international system.

Following the hypotheses developed in chapter 3, I identify here three features of the British system of environmental regulation: its highly decentralized structure, its closed policy community, and its highly flexible mode of policy implementation. I argue that these have fostered a high propensity to engage in the importation of hazardous wastes by playing a key role in affecting the following:

• the willingness and ability of the British waste disposal industry to import wastes,
• the continuance of waste importation, despite public opposition and the high degree of environmental risk posed to the population by highly toxic waste imports and poor quality disposal facilities,
• the trajectory of government policy toward waste importation—in particular, its inability to effectively implement the ban on waste imports it announced in 1991.

In the following sections of this chapter, I develop this institutional argument in detail, identifying and describing these features of Britain's regulatory system, then tracing out their effects on the behavior of industry, societal actors, and government officials. I then compare this explanation to the two alternative explanations, the first based on incentives for state actors to allow waste importation (economic-nationalist), and the second based on comparative advantage in waste disposal services. The chapter covers primarily the period between 1988 and 1996, the years for which waste importation data is available, and up to the point when the government invested the new Environment Agency with new powers.

Several themes run through the chapter. The first is the problem of managing environmental problems with both local and international characteristics in a unitary state with implementation powers delegated to a set of local authorities. In contrast to the clearly defined power-sharing arrangements exhibited by most federal states, this sort of system weakens effective monitoring and control of environmental policies—the principal-agent problem discussed in chapter 3. Another related theme is the relative impotence of the central government to implement desired policy changes and international commitments at the specific level of waste trade management, and their inability to effect broader institutional change. I conclude that, with respect to environmental policy at least, Great Britain, contrary to its reputation, is best characterized as a weak state. Finally, I briefly sum up how Britain's regulatory system is adapting to the impending pressures of the twenty-first century. The 1990s have been a turbulent time for British environmental policy, buffeted by demands from within for policy change and by international pressures from the EU. Conflicting interests often failed to find a voice in a political system caught up in the decline of the long-serving Conservative government and the ambitious agenda of Tony Blair's New Labour government. Therefore the pace of reform has been slow and much confusion remains: cornerstone legislation, such as the Environment Act of 1990, has been enacted but has taken years to implement, and hazardous waste importation figures have not yet been significantly affected. However, the direction of policy change in Britain, and the road it seems to have chosen, marks a considerable philosophical departure from earlier institutional rules and configurations. Its ultimate success depends not only on the adaptability of the environmental policy sector, but also on the extent to which the political system as a whole opens up to a wider range of inputs.

Dependent Variable: Waste Importation in Britain

Waste importation in Great Britain, first recorded in the mid-1970s, jumped sharply in the 1980s.[2] "In 1981, special waste imports amounted to less than 4,000 tons, but by fiscal year 1987/88, 80,000 tons of special wastes were being imported into the country" (Montgomery 1992:201; see also Allen 1992:167).[3] Britain's waste exports for disposal purposes, on the other hand, are minuscule.[4] Figures released by the Department of the Environment in 1995 reveal a steady trend in terms of growth of waste importation for final disposal between 1988 (when official reporting began) and 1994, as shown in table 4.1.[5] Table 4.2 shows imports of wastes listed as hazardous or controlled under Annex I of the Basel Convention to the United Kingdom in 1995 and 1996. Different hazardous waste categories were adopted under new EU regulations; therefore they are not exactly comparable to earlier figures.[6] These figures stand in stark contrast to the levels of net imports or exports of hazardous wastes displayed by Britain's OECD partners, as shown in tables 4.3 and 4.4.

Britain's major trading partners are its fellow European nations, with around 95 percent of its waste imports in 1993–94 coming from seven

Table 4.1
Imports of hazardous wastes to the United Kingdom, 1988/9–1993/4 (in tonnes)

	1988/9	1989/90	1990/1	1991/2	1992/3	1993/4
England and Wales	40,000	31,918	44,335	46,903	45,904	66,468
Scotland	27	0	182	17	164	40
N. Ireland	0	0	0	0	950	1,485
Total	40,027	31,918	44,517	46,920	47,018	67,993

Source: DoE/HMIP; SOEnD; DoE (NI), 1995. See http://www.environment.detr.gov.uk/des20/chapter7/index.htm for the 1998 Report.

EU member states: Germany, Ireland, the Netherlands, Luxembourg, Italy, Spain, and Belgium (as shown in table 4.5). According to ENDS, "barely 1,000 tonnes of the 179,000 tonnes of hazardous waste imported into England and Wales in the five years to 1992/3 came from developing countries and eastern Europe" (ENDS Report 233, June 1994:28).[7]

In terms of types of wastes imported, in 1992–93 seven different types of waste—fly ash/slag, filtration products, chlorinated solvents, PCBs, pesticides, "mixed" wastes, and sulfuric acid—dominated the trade, making up 70 percent of total waste imports during that period (ENDS Report 218, March 1993:24). PCBs are considered one of the most toxic and long-lived types of wastes in existence, and the others rank fairly high on lists of dangerous wastes.[8]

In sum, the above figures show a story of increased waste importation to Great Britain, a practice that began to be reported in the early-1980s and continued increasing into the 1990s. They also show that Britain not only imports its wastes primarily from its partners in Western Europe, but that it has consistently held its position as one of Europe's largest importers of hazardous wastes.[9]

Britain's System of Environmental Regulation

From the Nineteenth Century to the 1990s: A Historical Overview

As befits the earliest industrializing nation, Britain was the first country to enact pollution control measures, starting with the establishment of the Alkali Inspectorate in 1863 (Vogel 1986:31).[10] Since then, however,

Table 4.2
Imports of waste to the United Kingdom, 1995 and 1996 (in tonnes)[1]

Wastes consisting of, or contaminated with[2]	1995	1996
Health care wastes	196	536
Fine chemicals and biocides	9,861	14,270
Organic solvents	2,145	6,941
Inorganic and organic cyanides	0	0
Mineral oils, oil/water and hydrocarbon/water mixtures	1,986	131
PCBs and PCB contaminated items	1,487	1,991
Tarry residues from refining, distillation, and pryrolitic treatment	100	93
Inks, paints, dyes, and pigments	605	1,142
Polymeric materials and precursors	328	1,242
Photographic chemicals and materials	5,503	5,948
Wastes from the treatment of metals and plastics	95	1,261
Residues from industrial waste disposal	5,338	11,544
Heavy metals and their compounds	1,734	17,264
Arsenic and antimony and their compounds	0	8
Inorganic fluorine compounds	0	0
Acids and bases in solid form or solution	2,066	999
Phenols, ethers and organohalons not elsewhere specified	41	112
Wastes collected from households	0	0
Residues from incineration of household wastes	3,013	6,513
Unclassified[3]	15,950	11
England and Wales	50,449	70,006
Scotland	4,312	5,129
Northern Ireland	0	1,002
United Kingdom	54,761	76,137

Source: U.K. Department of Environment, Transport, and the Regions, *Digest of Environmental Statistics* No. 20 (1998).
Notes:
1. Imports of wastes under the Transfrontier Shipments of Waste Regulations 1994 between January 1, 1995 and December 31, 1996. 2. As defined by the categories of wastes in Annex 1 of the Basel Convention. 3. Used where the description of the waste was insufficiently precise to permit allocation to one of the Basel Convention categories.

Table 4.3
Net imports (exports) of hazardous wastes for selected OECD countries, 1989, ranked by net import level

	Imports (tonnes)	Exports (tonnes)	Net imports (exports)
United Kingdom	40,740	0	40,740
Denmark	26,842	8,978	17,864
Japan	5,125	40	5,085
New Zealand	0	200	(200)
Australia	0	500	(500)
Norway	0	8,078	(8,078)
Italy	0	10,800	(10,800)
Sweden	33,863	45,012	(11,149)
Austria	50,981	86,773	(35,792)
Netherlands	88,400	188,250	(99,850)
Switzerland	7,684	108,345	(100,661)
Germany	45,312	990,933	(945,621)

Table 4.4
Net imports (exports) of hazardous wastes for selected OECD countries, 1991, in rank order

	Imports (tonnes)	Exports (tonnes)	Net imports (exports), (tonnes)
France	636,647	21,126	615,521
United Kingdom	46,714	525	46,189
Austria	111,595	82,129	29,466
Japan	397	0	397
New Zealand	0	21	(21)
Australia	0	3,200	(3,200)
Denmark	15,200	21,758	(6,558)
Italy	0	13,018	(13,018)
Norway	2,415	14,636	(12,221)
Sweden	34,195	63,801	(29,606)
Netherlands	107,251	189,707	(82,456)
Switzerland	6,416	126,564	(120,148)
Germany	141,660	396,667	(225,007)

Sources: OECD 1993a and 1994a. Figures for the U.K. and Germany only include wastes destined for final disposal. Data on France for 1989 was unavailable. Figures for Japan in table 4.3 are from 1990.

Table 4.5
Imports of hazardous waste to England and Wales by country of origin, 1993–94

Country of origin	Amount of waste (tonnes)
Germany	32,644
Ireland	10,806
Netherlands	8,088
Luxembourg	6,494
Italy	2,495
Spain	1,300
Belgium	1,055
Switzerland	710
Portugal	573
France	344
Venezuela	287
Brazil	277
Sweden	263
Denmark	157
Korea	125
Greece	112
Austria	107
Other	631
Total	66,468

Source: Department of the Environment, London, 1995.

the environment has followed a roller coaster path as a policy priority. Flurries of legislation have been followed by years of inactivity, and only in the 1990s has any concerted effort been made on the part of the government and bureaucracy to formulate a coherent, integrated environmental policy (Garner 1996; Lowe and Flynn 1989). Policy development has been more reactive than proactive, responding to various crises—such as the 1952 London Smog, which gave rise to the 1956 Clean Air Act—or to international pressures. The 1974 Control of Pollution Act (COPA), the first piece of environmental legislation to combine crossmedia pollution under the same legislative framework, was enacted following the 1972 Stockholm Conference on the Global Environment. Britain's particular system of environmental regulation has proven remarkably durable over the years in terms of its overall style and structure, particularly in the face of external and internal pressures for change.[11]

As the century closes, Britain is moving toward an integrated system of pollution control (IPC).[12] This means that environmental functions will be gathered together under a single, central administrative roof, and that policies will have to take into account the crossmedia effects of environmental degradation. The impetus for these changes began in 1976, when the Report of the Royal Commission on Environmental Pollution "argued that the problem of cross-media transfers was the most important challenge that the regulation of pollution faced" and recommended moving "from a discharge control regime based on the idea of single-medium control to one based on the idea of multi-media control," or IPC (Weale 1996:114).

These ideas resurfaced during the final years of Margaret Thatcher's government, which—contrary to what might be expected—did not mark a rolling back of the frontiers of environmental regulation, at least in the realm of pollution control. In fact the period witnessed a spurt of institutional activity and the creation of a number of new environmental agencies, as well as the inception of the 1990 Environmental Protection Act, the keystone of subsequent reform. R.A.W. Rhodes argues that the British government, while seeking to minimize the role of the state in the economy, at the same time sought to maintain and strengthen its regulatory capacity, primarily to monitor the new private sector monopolies. This lead to the establishment of multiple new agencies, not only in the environmental sector but also across other policy areas: hence, "re-regulation" rather than deregulation (Rhodes 1994; 1996).[13] At the same time, however, the Thatcher regime facilitated and encouraged the sort of competitive and outward looking market economy conducive to the continued and expanded import of wastes from abroad.

Hazardous waste legislation has been in place in the U.K. since 1972, when the first legislation—the Deposit of Toxic Wastes Act—was enacted in a record ten days following the discovery of drums containing extremely toxic wastes in a deserted industrial yard being used by children as a playground. Further legislation includes COPA and its subsequent amendments, most importantly the 1980 Control of Pollution (Special Waste) Regulations, which focused on the proper handling, transportation, and disposal of "special" wastes.[14] Later in this section I discuss in more depth subsequent changes to British hazardous waste management practices under the 1990 Environmental Protection Act and related legislation. Hazardous waste regulation in Britain falls almost entirely under the rubric of pollution control regulation. This is in contrast to conservation and development policy, which concerns land use and planning, and the conservation of the national heritage (both natural and historical).[15] Traditionally in Britain the latter has dominated the former as a policy priority.[16] Correspondingly, Britain has—or had, at least until the late 1980s—a reputation for being a laggard when it comes to international environmental issues, especially those it is accused of inflicting on its European neighbors.

Structure of Hazardous Waste Regulation in Britain

Evolution of Britain's Structure of Environmental Regulation The evolution of Britain's environmental agencies has been somewhat haphazard: "in terms of structure, powers of control have been (and in the main still are) devolved among a confusing skein of regulatory authorities (local, regional, and national), implementing and enforcing a concretion of statues and common law" (Jordan 1993:408). Pollution control in Britain has traditionally been dealt with by medium (air, soil, or water) or by substance (Garner 1996:97), and at the national level several different ministries and associated agencies have environmental responsibilities. These include the Ministry for Agriculture, Fisheries and Food (MAFF), the Department of Industry, and the Department of the Environment, established in 1970 and subsequently renamed the Department of Environment, Transport, and the Regions. However, only about 10 percent of the latter's staff are directly engaged in environmental protection (McCormick 1993:270; OECD 1994b); its main environmental duties concern the oversight of the various agencies set up during the 1970s to protect different aspects of the environment: the Industrial Air Pollution Inspectorate, the Wastes Inspectorate, the Radioactive Substances Inspectorate, and the Water Quality Inspectorate (Garner 1996:97).[17]

During the 1980s, steps were taken to amalgamate the different inspectorates into one body, Her Majesty's Inspectorate of Pollution (HMIP), established in 1987. While this represents the first step toward an integrated system of pollution control, the change was merely cosmetic until the early 1990s, when HMIP was given the means to implement crossmedia policies (Weale 1992:105–107). This legislative ability was provided under the Environmental Protection Act of 1990; however, it was another three years before IPC authorizations began (Jordan 1993:409).[18] In 1991, Prime Minister Major announced plans for an Environmental Protection Agency (EPA), which would combine the functions of HMIP and the National Rivers Authority (NRA) in a single body. This proposal initiated several years of discussion and planning, and the creation of the EPA was postponed on numerous occasions. It finally came into being on April 1, 1996, established as an agency independent of direct government control.

At the subnational level, local authorities have responsibility for controlling sources of pollution not covered by the above agencies and are also responsible for implementing legislative frameworks—often taking the form of extremely loose sets of guidelines—established by the central agencies (Vogel 1986).[19] These authorities vary substantially in their abilities to effectively implement environmental policies. Related as parts of a unitary rather than federal political system, local authorities in Britain are completely reliant on central government, which sets out their duties and controls their sources of funding. They also do not have an independent base of political legitimacy and representation at the national level in the way that the constituent units of federal states, such as the German Länder, do. Furthermore, "unlike the German Länder, the political actors at the subnational level—the local authorities—are not systematically involved (institutionalized) in the decision-making process at the central level of government" (Héritier et al. 1994:158).[20]

The implications of this particular pattern of responsibility and authority were seen dramatically during the Thatcher years, when conflict between central and local government became a hallmark of the political scene. The failed implementation of the poll tax, which sought to radically restructure, and indeed undercut, the funding base of local authorities, became the main symbol of these years of political strife.[21]

Central government, under the Local Government Planning and Land Act (1980), took over more control of local government funding and sought to promote those elements of local government more amenable to its overall policy plans. It also reshaped the relationship between local authorities and their communities by introducing the compulsory sale of public housing and the competitive tendering, or contracting out, of local services such as road maintenance and refuse collection.[22] The environment provided an issue over which local authorities were able to defend their role in local affairs and augment their (admittedly low) legitimacy (Ward 1993:465–468). However, the overall climate of cutbacks and uncertainty severely reduced their ability to act effectively on issues of environmental protection, even though their overall duties remained unchanged.[23]

Structure of Hazardous Waste Regulation The central philosophy behind Britain's structure of hazardous waste regulation has been the view that wastes, regardless of point of origin, are purely local in their effects. Therefore they should be dealt with at a primarily local level, despite the technical complexity of waste management and the increased role of the EC/EU in this area (Barnes 1997:170). This view is spelled out in the Report of the House of Lords Select Committee on Science and Technology, on Hazardous Waste Disposal: "[b]ecause the effects of pollution are usually experienced first within the confines of particular localities, one of the principles followed by successive Governments has been that the primary responsibility for dealing with pollution problems should rest, as far as is practicable, with authorities operating at a local or regional level. Thus, central Government lays down the statutory framework for pollution control, but implementation is delegated to a large extent to local level" (House of Lords Select Committee, 1981, Vol. II:92, quoted in Mädel and Wynne 1987:202). Alternatively, as one government minister put it: "local knowledge and the capacity for a fast response which lies with local authorities could be put at risk by a centralized system."[24]

These views generated one of the most highly complex, confusing, and decentralized waste management structures in the world. Waste regulation responsibilities are divided not only among different levels of government but also across literally hundreds of regulatory agencies that, following legislative innovation in the early-1990s, remain in flux. The same views also underlie the official government policy of treating imported wastes as equivalent to domestic wastes for disposal purposes.

Britain's hazardous waste disposal industry is privately owned and competitively structured. Wastes are shipped to private firms that specialize in their disposal, and the United Kingdom is unique in Western Europe in relying on private venture capital for all of its specialist facilities (ENDS Report 125, 1985:11, citing the First Report of the Hazardous Waste Inspectorate). This structure creates the need for an extra tier of regulation to control the social costs of waste disposal (externality effects), which private firms are unlikely to take into account.[25]

British hazardous waste regulation rests on the regulation of this private industry by local and regional waste disposal authorities "through a combination of site-licensing, inspection, waste registration and legal enforcement mechanisms" (Mädel and Wynne 1987:195). This framework came into existence in 1972 under the Deposit of Poisonous Wastes Act and was subsequently institutionalized through COPA and the 1980 Control of Pollution (Special Waste) Regulations, only changing in 1996 with the creation of the Environmental Protection Agency.

Many waste disposal authorities tied to town, city, and county councils have traditionally borne responsibility for waste regulation: one estimate identifies approximately 193 authorities with waste-related responsibilities as of 1994 (Barnes 1997:179).[26] Under the 1990 Environmental Protection Act, three types of authorities have a role in waste management (Barnes 1997:176). Waste Regulation Authorities (WRAs), tied to county or district councils, administer hazardous waste regulations. Waste Collection Authorities (WCAs) primarily collect household waste. Waste Disposal Authorities (WDAs), previously the main regulatory bodies, now coordinate private disposal contracts and regulate disposal sites.[27] Centralized monitoring did not come into being until 1983, when, in the light of severe inadequacies arising in certain authorities, the government established the Hazardous Wastes Inspectorate. The HWI was subsequently incorporated into HMIP as the Controlled Waste Inspectorate, whose primary duty is to monitor and report on the activities of the WDAs.[28]

Given the official government policy that for the purposes of disposal, wastes imported from abroad are treated no differently from wastes produced at home, wastes imported to the United Kingdom are not policed at the national level. Rather the WDAs, now the WRAs, "receive and process pre-notifications of hazardous waste imports" (ENDS Report 187, August, 1990:18). This is important for a number of reasons. First, responsibility for what is essentially a foreign policy issue has been abrogated to local authorities that have no control over other areas of foreign policy. Second, imported wastes are treated as having arisen at their point of entry into the United Kingdom (indeed, at their point of entry into the local authority's territory), creating a whole range of problems should the need for assigning transboundary liability arise. While movements of waste are subject to notification and transportation regulations, in the absence of authoritative international controls over these, it is hard to imagine, for example, a WRA in Manchester successfully prosecuting a firm in Germany in the event of some accident involving imported wastes. Waste movements across local authority boundaries within the U.K. is also common, with several regions—notably Scotland and Wales—exporting 50–100 percent of their wastes, most of which are received by Bedfordshire, Greater Manchester, Essex, Cheshire, and the West Midlands (ENDS Report 223, August 1993:30).

Local regulatory authorities are therefore in a double-bind with respect to hazardous wastes: "whereas the public [local] authorities have a responsibility to collect and manage *domestic* wastes right from their point of production, they are 'outsiders' to most if not all of the *hazardous industrial* waste life cycle and do not control what leaves or enters their area" (Mädel and Wynne 1987:233; emphasis added). Waste regulation is focused almost entirely on waste disposal, giving regulatory authorities little control over waste content in terms of hazardous elements.

A variety of factors have combined to undercut the ability (and even desire) of local government to carry out its job in this area. Local government financing is extremely precarious. Local authorities have been under severe financial strain since the early 1980s, when their powers were severely circumscribed and their revenue bases slashed by the Thatcher government. Furthermore, most (some would argue all) local authorities have insufficient capacity to control the actions of the firms under their jurisdiction or to influence central government policy (Marshall 1998:439). Fulfilling their main task—ensuring the safe disposal of hazardous wastes—is enough of a challenge without having to worry about whether the wastes are foreign or domestic in origin (Mädel and Wynne 1987:236).[29] Waste disposal companies are an important source of revenue for these authorities, providing local employment and expanding tax revenues; hence, local authorities make the effort to attract companies to their areas. Concomitantly, disposal companies have turned to waste importation as a way of maintaining productivity and boosting profits, a move facilitated by the policy-dictated indifference of local authorities to the point of origin of wastes.

Britain's Style of Environmental Regulation

Policy Access: A Closed Policy Community In terms of the continuum identified in chapter 3, Great Britain has an exclusive environmental policy community. State and industry representatives cooperate in the formation and implementation of environmental policy—a process that excludes societal interests.[30] The closed policy process is accentuated by the absence of broader channels of access; in Kitschelt's terms, Britain also has a closed political opportunity structure (Kitschelt 1986). As one analysis puts it, "environmental policy-making, where it impinges on major economic interests like pollution control, has traditionally taken place in closed policy communities, in which producer interests—industrialists, trade associations, farmers and landowners—are heavily represented" (Carter and Lowe 1994:266). According to another, "because many policy-makers think of themselves as custodians of the public interest, and feel they understand the best interests of the public with minimal reference to the public itself, environmental policy in Britain tends to be made in closed policy communities" (McCormick 1993:269). Thus the relationship between government and the regulated industries has traditionally been very close. Channels of access to the policy process are typically informal, and the system operates in such a way that major societal and environmental groups are excluded from this process.

Environmental policy in Britain is made largely within government agencies (the relevant branches of the civil service, special committees, and local government agencies) through a process of consultation and cooperation. Experts and industry representatives are called in to testify before policy committees or are consulted informally, and legislation is then put before the House of Commons and the House of Lords. However, unlike Germany or the United States, Britain has no separate process of judicial review, which effectively cuts off one of the main access channels used by environmental interests in those countries (Héritier et al. 1994:166). Furthermore, owing to the technical and scientific nature of many environmental policy decisions, and hence the need for specialized knowledge in these areas, key influence within the policy community is further limited to those actors with the required expertise (Weale 1992a; Enloe 1975).[31] The natural bent of the British political system toward secrecy and the restriction of information from the public on a "need to know" basis have also reinforced this policy style (Brickman, Jasanoff, and Ilgen 1985:44). However, in recent years the government has in fact made an effort to increase accountability: the widely distributed, and highly critical, reports of the Hazardous Waste Inspectorate are a case in point.

The absence of formal and easily called upon processes of judicial review of state action is the first indicator that Britain's broader political opportunity structure is also relatively closed. In terms of parliamentary representation of environmental constituencies, Britain is one of the few countries in the world with a first-past-the-post electoral system. Under this system, Britain is divided into roughly 650 territorial single-member constituencies. In each of these, the party representative with the most votes will win; the party with the highest number of overall seats wins the election. This system, unlike systems of proportional representation that predominate elsewhere in Europe, is systematically biased toward two large, inclusive parties that are able to organize at a local level in a wide number of areas and to appeal to a broad range of voters. This works to exclude smaller, primarily issue-based parties such as the Greens, who appeal to territorially scattered voters (Bogdanor 1983). The two major parties, Conservative and Labour, neither of which rely on environmental interests for their electoral base, have therefore dominated Parliament.[32] Finally, Britain is a unitary state, with no separation of powers as exists in the United States. Even though environmental regulatory powers are decentralized, political power, and therefore lobbying efforts, are concentrated in Westminster. Far fewer access points for public opinion and interest groups to influence important policy decisions exist than in federal states such as Germany, the United States, or Australia.

Flexible Mode of Policy Implementation The United Kingdom has a flexible mode of policy implementation: "the British make much less use of legally enforceable environmental quality or emissions standards than does any other industrial society. They attempt to tailor pollution-control requirements to meet the particular circumstances of each individual polluter and the surrounding environment" (Vogel 1986:75–6; see also Lowe and Flynn 1989:257 and Brickman, Jasanoff, and Ilgen 1985:306). Policy implementation and evaluation has traditionally been based on the "best practicable means" (BPM) criterion: rather than firms having to meet nationally imposed pollution standards, they simply have to ensure they are doing the best they can, on an individual basis, to protect the environment, given the means at their disposal.[33] Skea and Smith report that BPM was considered an "elastic band" on the part of the agencies (Skea and Smith 1998:267). Hence the British system relies on voluntary compliance on the part of industry. This is compounded by the official definition of "special wastes" that is applied in the United Kingdom (see note 14, this chapter); much more restrictive than those used in other countries, it enables waste importers to import wastes categorized as hazardous elsewhere under less stringent requirements (Allen 1992:171).

As Weale notes, this marks a major philosophical difference between Britain and countries such as Germany, which rely on rigid, nationally imposed standards for environmental protection: "From the point of view of German policy, the British BPEO [best practicable environmental option] looks often as though it confuses the claim that there is no evidence of environmental damage with the claim that there is evidence that there is no environmental damage. From the British point of view, the German principle of precaution looks like assuming that there is a risk and being prepared to control it, even though there is no evidence to warrant the conclusion" (Weale 1995:21). Mädel and Wynne relate this philosophy to the U.K.'s broader regulatory culture: "The U.K.'s technical imprecision and uncertainty in regulation is not merely a result of its strong institutional traditions of mutual confidence and trust; it is a *necessary* dimension of wider U.K. processes, especially intragovernment and industry-government relationships. This cultural style does not merely allow, but *needs* technical uncertainty as a currency of informal institutional interaction, demonstration of good faith, and willingness to compromise—the hallmarks of U.K. regulation" (Mädel and Wynne 1987:234; emphasis in original).

Thus the British mode of environmental policy implementation gives private firms, be they domestic manufacturers or waste importers, considerable leeway in setting their own goals with respect to environmental protection, and firms are rarely prosecuted for infringement of regulations.[34] This is not to imply that environmental quality is necessarily lower in Britain as a result; however, it does mean that certain activities—such as waste importation—come under less automatic scrutiny than in other countries, unless a case arises where definite environmental damage occurs.

Effects of the Regulatory System on Waste Importation

Effects on Industry Behavior: Enabling and Facilitating Waste Importation

Britain's hazardous waste disposal industry turned to waste importation from abroad as a means of generating revenues during the 1980s. In doing this, it was able to take advantage of industry's "special relationship" with government, as well as a regulatory climate that enabled it to engage in waste importation and gave it the flexibility to set its own goals and priorities without much outside consultation.

Britain's waste disposal industry is the exemplar "private competitive" model of industry structure (Linnerooth and Davis 1987; see chapter 3). The vast majority of waste disposal plants are owned and operated by companies in the private sector that compete for disposal contracts. Integrated disposal facilities, which combine a range of different disposal options and processes in one plant, do not exist in

Britain, and government subsidization of hazardous waste disposal facilities is very limited and getting lower (ENDS Report 202, November 1991:27). "The private sector accounts for 98% of all industrial waste disposal, half of this being conducted "in-house" at the site of production.... The waste T&D [treatment and disposal] industry is in the hands of several large and many small companies, whose trade association is the National Association of Waste Disposal Contractors (NAWDC), which includes about 75% of the T&D industrial sector" (Mädel and Wynne 1987:207).

In 1992, around 2000 firms existed in the waste disposal sector. Over the past thirty years, the industry has evolved "from a disparate collection of localized small-scale operators to a coherent multi-million pound industry" (Cooke and Chapple 1996:11), now worth an estimated £3 billion.[35] Britain's waste disposal industry has been dominated by differing configurations of five big firms.[36] Furthermore, foreign companies —notably the two leading North American firms, Waste Management and Browning-Ferris industries, and French firms such as SITA, Compagnie Générale des Eaux, and Saur—have set up British subsidiaries (Cooke and Chapple 1996:14).

In the 1980s, according to Greenpeace, most hazardous waste imports to Great Britain went to a few large companies: Leigh Environmental—based in the West Midlands (26%); ReChem International—based in Pontypool and Southhampton (23%); Lanstar Wimpey Waste, Leigh Environmental, and PJ Collier—Manchester (39%); Cory Waste Management, Cleanaway, and Max Recovery (Vallette and Spalding 1990: 354). ReChem and Lanstar subsequently came to dominate the market, the former importing around 10–12,000 tonnes of waste annually for incineration, although its position was challenged in the latter part of the decade by Leigh (Allen 1992:189; ENDS Report 228, January 1994:34).[37]

In this context it is not hard to identify why Britain's waste disposal industry wants to engage in waste importation. Private firms operating in an increasingly globalized and competitive, but at the same time uncertain, market naturally seek out all possible sources of profit available to them, and it is estimated that imported wastes earn roughly £1,000 per tonne.[38] The economic climate in the 1980s reinforced this tendency. The Thatcher government was engaged in a fundamental shakeup of British industry, decreasing the level of state involvement in economic activity and increasing competitiveness. This period also marked the beginnings of the process of European integration that culminated in the Single European Act of 1987 and the Treaty on West European Union. This expansion of trade within the European Community, at this point in the absence of related environmental policies, greatly facilitated the movement of wastes across national frontiers (Laurence and Wynne 1989).

As Vogel notes, relations between government and industry in the making of British environmental policy have evolved over time into a tightly knit and friendly relationship, "predicated on a high degree of cooperation between the regulators and the regulated" (Vogel 1986:83). In contrast, in the United States government-business relations with respect to environmental regulation have become more adversarial than cooperative; significantly, Vogel argues, U.S. policies reflect a much higher degree of risk aversity than those adopted in Britain (ibid.:253). This observation holds equally true for the relationship between British policy makers and the waste disposal industry: "[t]he most important observation about the U.K. system [of hazardous waste regulation] is the remarkably strong dependence upon informal trust and collaboration among the various institutional actors, at all levels" (Mädel and Wynne 1987:234).

When debates about waste importation began in the early-1980s, waste disposal companies, represented by the National Association of Waste Disposal Contractors (the NAWDC, now the Environmental Services Association), played a key role in policy decisions. Montgomery cites the involvement of the waste disposal industry in policy negotiations in 1981, when NAWDC representatives testified in favor of allowing continued waste imports before a House of Lords Review Committee.[39] In contrast, "of the 641-page transcript of written and oral evidence submitted to the Select Committee, only six pages were submitted by environmental groups and none by citizen activists" (Piasecki and Brooks 1987:194). This involvement has continued to the present day. The industry's primary concerns are both to maintain waste importation practices and to restrict the ability of smaller "cowboy" firms from gaining a foothold in the market.[40] The most notable victory in recent years was the success ReChem achieved in delaying a proposed ban on waste importation by three years (ENDS Report 233, June 1994:28).

In sum, a picture emerges of an industry operating under regulatory conditions extremely favorable for waste importation. Large private firms enjoying a favorable relationship with government actors are subject to little legislative scrutiny, particularly in terms of where they obtain the wastes they process. They exist in a highly competitive market environment characterized by potentially extremely high profit margins and by economic pressures and uncertainties both at home and abroad. The absence of legislative measures regarding the waste trade at the outset enabled firms to begin importing wastes; they have subsequently been able to use their privileged position in the policy process to continue these practices. I examine the attractiveness of the U.K. waste industry to foreign waste producers in my later assessment of the U.K.'s comparative advantage in waste disposal.

On the Periphery: Environmental Groups and Public Opinion

Environmental activism has a long tradition in Britain. However, unlike most other industrialized countries, the environmental movement has few channels of access to the policy process, and the environmental activism that exists is not often effectively translated into policy outputs, as one might expect in a pluralist democracy. In this section I describe Britain's environmental movement and public opinion with respect to the waste trade and analyze the institutional mechanisms that have tended to filter them out of the policy process.

There is little doubt that strong opposition to the waste trade exists in Great Britain. Allen (1992) discusses several well-documented cases of organized local opposition to waste disposal or planned facilities in both Britain and Ireland, and larger environmental groups such as Greenpeace and Friends of the Earth regularly lobby against the trade at the national level. In 1991 the British Medical Association published a well-publicized report calling for a ban on waste imports (Allen 1992:217), and in August of 1994 the *Daily Mirror*, one of Britain's tabloid newspapers, ran a front page story under the headline "Ship of

Doom" about the proposed off-loading in Liverpool of a cargo of contaminated waste from Germany. Similar stories, as well as extensive television coverage, ran during the Karin B. affair in 1988.[41] These pieces took on a distinctively nationalist tone in protesting the importation of wastes from other developed countries, especially Germany. Concern over hazardous waste disposal ranks high in a 1996 survey of public attitudes toward the environment: 60 percent of respondents said they were "very worried" about dangerous wastes compared with other environmental problems.[42] Nor is it the case that hazardous waste disposal facilities are located in poor or minority communities that lack the strength or political clout to organize against them, as has been argued is the case in the United States. Britain's waste facilities are located across a broad cross-section of socioeconomic regions, usually densely populated, and figures on domestic waste movements show that wastes tend to move out of the politically marginalized regions, such as Scotland, for disposal in other parts of the country.

Britain's environmental movement differs in several important respects from those found in many other industrialized countries. First, the strongest and most well-established part of Britain's environmental movement is concerned more with issues of conservation and heritage preservation—more often the responsibility of public organizations in other countries (Vogel 1986:49)—than with pollution control.[43] Britain has virtually no tradition of an activist deep ecology movement (Rootes 1995).

Groups that deal with issues of pollution control, and hazardous wastes in particular, are in contrast relatively new on the political scene, and are not as well established as the conservation organizations.[44] The most organized and widely supported of these are Greenpeace and Friends of the Earth (FoE).[45] The latter is probably the only group of its kind with insider status in the policy process; in the early-1980s it shifted from a more radical stance to a consultative one in order to enhance relations with particular government agencies (Vogel 1986:51). In particular, it targeted local authorities, drawing up plans for implementing environmentally sound policies in their own areas (Grant 1995:94; Ward 1993). Even so, "both sides are still wary of each other, and tensions do remain" (Ward 1993:472), and FoE has often been accused by its partners in the Green Movement of having been co-opted by government or bureaucratic interests (Vogel 1986:51). Vogel goes on to argue that FoE has only been allowed to occupy a "peripheral" insider position, with access to government agencies with "limited power and small budgets" (ibid.:52); larger ministries that deal with pollution control, such as the Department of Industry and MAFF, remain insulated from such contacts.[46] Greenpeace is well known for its activism regarding toxic waste issues; however, Britain does not have the same established networks of toxic waste activist groups that exist in the U.S. or other European countries (Piasecki and Brooks 1987:193–5). According to Allen, neither Greenpeace nor FoE have the local community orientation to really push these issues, and where priorities conflict they are likely to favor international causes over local concerns, also considered an easier task than tackling the "powers that be" in Whitehall (Allen 1992:221).

Public opinion, too, appears to have had little direct influence on policymaking. Vogel argues that environmental consciousness is relatively low, in the sense that there is a fairly high degree of acquiescence to and satisfaction with government policies and their outcomes—for example, the astonishing transformation of air quality in Britain's inner cities in the period since the 1956 Clean Air Act. This, however, is not altogether an accurate picture. Many controversies—including the export of live animals for slaughter, attempts to destroy historic areas of woodland to make way for motorways, the wave of public outrage in the wake of the "Mad Cow" debacle of 1996, and the proposed introduction of genetically modified organisms into U.K. agriculture in 1998 and 1999—have led to highly publicized protests. These protests, though small in numbers, have involved a broad cross section of the public and have won a high media profile.[47] More generally, studies show a large increase in public awareness of environmental issues in Britain: "in the UK the proportion of people viewing the environment as an urgent problem rose from 67 per cent in 1988 to 82 per cent in 1992. In Germany the rise was from 84 percent to 88 percent" (Peattie and Ringler 1994, citing a 1992 Eurobarometer study).[48] However, public debates, instead of occurring before the appropriate policy decisions were made, have more often been fought out, sometimes literally, "in front of the bulldozer." Several waste-related controversies—notably that surrounding the waste ship *Karin B.*—have emerged only subsequently to information being leaked to the popular press.

In general there is a widespread feeling among the public that they are excluded from the environmental policymaking process and that their concerns will not be accurately reflected in government policy.[49] Nor do they trust Whitehall to manage their concerns, although the belief in individual action is quite strong (Worcester 1997). The somewhat paradoxical picture of inactivity on the part of the majority coupled with the unconventional, even radical tactics followed by the more active environmentalists can be directly related to the absence of formalized institutional channels for public opinion to express itself in the policy process. Reasons for this include, as discussed above, the first past the post electoral system, which excludes direct participation at the national level by Britain's tiny Green Party, the concentration of policymaking powers behind closed doors in Whitehall and Westminster, and the absence of formal judicial review processes.[50] Conservative Party dominance in the 1980s further insulated the system from the claims of environmentalists (Boehmer-Christiansen and Skea 1991:281). During the high years of Conservative Party rule, the role that environmental groups had been playing in the policy process (for example, testifying to committees) was considerably curtailed, and several quasi-governmental advisory groups were axed (Lowe and Flynn 1989:266).

Britain's environmental movement is fragmented, and many of its tactics are at best tangential to the mainstream of British politics. Groups that mobilize around a single issue are unlikely to coalesce into a longer-term network, and even then it is rare for them to be incorporated into the Whitehall-Westminster nexus in the same way that business groups have been.

Effects on Government Actors: Words Louder than Actions

An examination of official policy and policy outcomes regarding waste importation demonstrates that government actors at both central and local levels are as much influenced by a country's regulatory system as they are by the societal actors the rules have been designed to govern. Britain's style and structure of environmental regulation are long-standing, and have evolved over the years, until very recently, without any significant break in practices and modes of operation. In this section I examine the evolution of policy at the national level toward waste importation and the factors constraining government actions in this area.

The official government position on the waste trade has shifted from "no policy" in the early-1980s to fairly strong support of waste importation to an announced ban in 1991 on the importation of hazardous wastes from other industrialized countries. This ban can be related to EC/EU policies regarding the need to achieve self-sufficiency in waste disposal and the emergence of the international regime governing the waste trade, which brought the trade under international legislative scrutiny for the first time. Perhaps more importantly, the extremely high levels of publicity given to the *Karin B.* and similar incidents also prompted this announcement. Even with this level of opposition, however, contradictions appear. For instance, when the *Karin B.* attempted to offload its cargo in Britain in 1988, the government withdrew its permission on the basis that the exact composition of the wastes had not been ascertained. According to Montgomery, it was this incident, along with the cargo of wastes found in the hull of the wrecked ferry *Herald of Free Enterprise* in 1987, that prompted the government to implement EC regulations on the provision of written consent and consignment notes (Montgomery 1992:211–218). However, shortly thereafter Environment Minister Virginia Bottomley confirmed the government's position, stating that "it remains [the government's] view that international trade in waste-disposal services can be economically and environmentally desirable—provided that it is properly controlled and monitored."[51]

However, the waste importation ban has not yet been implemented. As mentioned earlier, as late as 1994 ReChem managed to obtain a three-year grace period to phase out imports, and it was only in May 1996 that an official government policy document on managing the waste trade finally appeared.[52] To many, this is incomprehensible: it would seem that stopping shipments of hazardous wastes from abroad would be a relatively simple matter for customs officials to deal with, and the government has been accused of "ducking the issue" (ENDS

Report 231, April 1994:34). The answer, again, lies in the way that the regulatory system has constrained the government's freedom of action in policy implementation.

First, this procrastination can be related to the close business-government relationship on these issues to the exclusion of broader societal groups. The Environmental Data Service (ENDS) reports that ReChem and Cleanaway have "lobbied hard" to achieve concessions with respect to imports, and have indeed succeeded in this.[53] Hence the privileged position of business interests shows up very clearly in the relevant policy debates. It is hardly even necessary to argue that this is a form of regulatory capture. Rather, patterns of interaction between government, business, and societal groups reflects a long-standing relationship that in turn has come to be reflected in the specific rules and practices that make up Britain's system of environmental regulation. ENDS also notes in "a bleak mid-term progress report" on the 1990 Environmental Protection White Paper that the government has been able to take advantage of its relative lack of accountability in failing to fulfill its pledges to implement environmental policy change (ENDS Report 232, May 1994:14–17).

Second, the decentralization of regulatory powers in this area has reduced not only the monitoring capabilities of central government, but also levels of cooperation and coordination over waste import controls. With local government holding the main responsibility for issuing waste importation permits as well as ensuring the adequate disposal of wastes within their jurisdiction, central government is several steps removed from "on the ground" management of imported hazardous wastes. Local authorities on the other hand, as argued above, have every incentive to allow the importation of wastes into their jurisdictions, be it from other parts of Britain or from abroad, a fact taken full advantage of by waste disposal companies.[54] It is also doubtful that these authorities have the ability to assess whether exporting countries do in fact have inadequate facilities for disposing of the wastes, as required under international conventions (ENDS Report 233, June 1994:29).

Hence the British government has been extremely slow to act on banning waste imports because it is itself constrained by its system of hazardous waste management and regulation. Its susceptibility to industry lobbying, in addition to the devolution of regulatory powers, means that however well intentioned the government might be, it continues to find it very difficult to implement the proposed policy changes in this area.

Alternative Explanations

State Calculations of the Relevant Costs and Benefits

Focusing exclusively on the financial incentives for government actors to allow waste importation, the economic-nationalist argument posits that Great Britain imports wastes because the rewards or benefits reaped from the trade outweigh the risks incurred by allowing such importation. This implies that state actors are able to act in a rational way in arriving at the net benefit-cost outcome. According to Montgomery, "the United Kingdom considered the environmental risks of importing hazardous waste, and weighed them against expected financial or economic benefits and determined that, in general, the benefits of importing hazardous waste outweighed the risks" (Montgomery 1992:221). Attempts to quantify the immediate financial gains from waste transactions tend to arrive at large positive sums. For example, one estimate puts revenues in 1987 at £1.2 billion (Montgomery 1990:316); according to another, "the value of toxic waste imports to Britain in 1988 was over £700 million" (Allen 1992:xiv). Furthermore, according to the British government the hazardous waste importation business is an important contributor to the country's balance of payments (Vallette and Spalding 1990:351; Allen 1992:172).

This argument is problematic on a number of grounds, both theoretical and empirical. Several issues arise immediately on examination of calculating risk in this situation. One, based on the analysis presented in chapter 2 and the evidence above on public opinion regarding the waste trade, is that societal evaluations of the risks associated with waste importation tend to be much higher than "expert" opinion. Rightly or wrongly, in Britain these evaluations are given little weight in the policy process, and the reasons for this involve the lack of public access to this process. Second, as I make clear in the next section of this chapter, the actual risks posed by the disposal of hazardous wastes in the U.K. are far from negligible given its reliance on landfill as the primary means of waste disposal, an opinion expressed by Britain's own Hazardous Waste Inspectorate. Britain already faces a huge bill for the cleanup of contaminated land, and it seems irrational to want to add to that total.[55] Finally, and this will become apparent when the British case is compared with that of other countries, this explanation does not provide any leverage in explaining why other countries do not try to reap similar rewards, or alternatively, why they arrive at different risk-benefit calculations.

On theoretical grounds, this argument is flawed in the sense that it relies on the assumption of the state as unitary (and autonomous) actor. In Montgomery's examination of the British case, for example, decisions regarding waste importation are assumed to be made by the national government. However, as the above arguments demonstrate, the fragmented regulatory structure that exists in the United Kingdom militates entirely against this assumption. Instead, a disparate group of regulatory authorities make these decisions, not only independent of one another but also in the absence of sufficient information on which to base their decisions. There is little doubt that the British government since 1979, famously accused by former Prime Minister Harold MacMillan of "selling off the family silver" in its privatization programs, is one that has valued economic benefits over environmental risk. The relevant issue is, however, why it has been able to do this and remain relatively immune from the pressures of public opinion.

Comparative Advantage in Waste Disposal

The second alternative explanation put forward in chapter 3 posits that Britain imports wastes because it has a comparative advantage in waste disposal, based on capacity and quality of waste disposal facilities. Countries such as Britain, the argument goes, offer high-quality disposal technologies or specialized facilities not available elsewhere, enabling the neutralization of the most harmful wastes in a way that poses minimal risk to the population. In the following sections I assess both aspects of this explanation and demonstrate that Britain's comparative advantage in waste disposal lies in terms of low-cost rather than high-quality facilities. This might explain why companies choose to export

Table 4.6
Waste disposal methods in the United Kingdom, early 1990s

	Controlled waste (140 million tonnes p.a.)	Special waste (2.5 million tonnes p.a.)
Landfill	70%	70%[1]
Incineration	5%	5%
Recycled/reused	25%	—
Sea dumping	—	10%
Physical or chemical treatment	—	15%

Source: OECD 1994b, figure 4.5, Department of the Environment.
Note:
1. Figure excludes residue of other treatments. Alternative figures, for 1992, put the amount of hazardous wastes receiving treatment at 8%, marine disposal 6%, incineration 2%, and landfill at 84%; from a report issued by Barclays de Zoete Wedd, released by the ICC Information Group, April 1, 1993 (ICC Report No. 048401).

their wastes to the United Kingdom, but it is a poor explanation of why government officials choose to issue import permits.

Disposal Technology Studies show that most of Britain's own and imported wastes are ultimately disposed of through landfill. One study puts this figure at 85% in 1985 (Forester and Skinner 1987), another at 70% in the early-1990s (OECD 1994b).[56] Table 4.6 gives a breakdown of Britain's disposal facilities. With respect to imported hazardous wastes, "about 47 per cent ... was physically or chemically treated before final disposal, and 42 per cent was incinerated" (OECD 1994b:73). Final disposal—for example, of incinerator ash—is almost always onto land or into water. These figures square roughly with the DoE figures for the ultimate destination of waste imports for 1993–4: of the 66,468 tonnes imported that year, 39,160 were incinerated, 19,711 were physically or chemically treated, and the remaining 7,597 tonnes received some other form of treatment.[57]

Two particular features of British landfill practices stand out. First, the majority of landfills in Britain are "dilute and disperse" sites, that is "[unlined] landfills that are built over permeable soils in order that their wastes can seep out of the site—on the assumption that they will be broken down and rendered harmless as they filter through the underlying soils" (Hildyard 1986:226). Second, Britain is the only country in Europe to employ (and officially endorse) the practice of co-disposal, whereby hazardous wastes are mixed with other types of wastes, usually in unknown quantities, following the principle that hazardous elements will be neutralized through reaction with elements present in the other wastes (Postel 1987:13; Piasecki and Davis 1987:193). According to one report, roughly 95% of liquid hazardous wastes are codisposed, 3% solidified, and 2% deposited in mineshafts.[58]

The British government has until recently defended the policy of relying on landfill as the primary means of waste disposal, even though the practice is coming increasingly under fire elsewhere. It argued that the practice was safe and used only for low-level toxic wastes. On the other hand, Sandra Postel quotes a 1986 report by the Hazardous Waste Inspectorate that "paints a grim picture of actual practices, and concluded: 'if we have avoided major problems with co-disposal landfill in the U.K., the Inspectorate considers that in some cases this is due more to luck than judgment'" (Postel 1987:13).[59] This view has also been echoed in a recent survey of Britain's landfill sites by one of the leading firms, U.K. Waste Management Ltd.[60] Safety questions have also arisen about the operation of British hazardous waste incinerators. In 1991 elevated levels of PCBs and dioxins were found in the air and soil around ReChem's Pontypool incinerator, the only facility in Britain capable of handling PCBs (ENDS Report 210, October 1991:12–13), and many claim that Britain's incinerators do not meet the standards imposed by most other countries (OECD 1994b).[61]

Disposal Capacity Figures on waste disposal capacity in Britain are available from a variety of sources, although none is completely up to date. Table 4.7 shows that according to the OECD and the DoE, there are roughly 7,600 licensed disposal facilities in the U.K. Of these, landfill, incineration, and treatment plants are reported to be used for the disposal of hazardous wastes; storage plants, too, are used to hold the wastes prior to final disposal. The table omits the other main source of waste disposal, ocean dumping. In terms of available capacity in tonnage, or volume, table 4.8 shows estimated disposal capacities in 1984.

Table 4.7
Disposal licenses in England and Wales, 1991–94

Type of disposal facility	Number of disposal licenses		
	1991–92	1992–93	1993–94
Landfill	4,228	4,077	3,435
Civic amenity[1]	589	676	764
Transfer station	1,453	1,712	1,976
Storage	344	355	370
Treatment	225	274	328
Incineration	234	151	214
Recycling/recovery	74	96	110
Scrapyard	713	1,041	1,670
Other[2]	11	12	38
Total	7,412	8,394	8,905

Source: Cooke and Chapple 1996:13; Department of the Environment 1995.
Notes:
1. Collection point for household waste. 2. For example, mineshafts. Note these numbers apply to all controlled wastes, not just special wastes.

Table 4.8
Data available on hazardous waste disposal capacities, Great Britain, 1984

Method	Capacity (1,000 tonnes)
Incineration (p.a.)	65
Chemical (p.a.)	300
Landfill (in 1984)	3,740
Sea dumping (in 1984)	260
Estimated total annual capacity	4,365

Source: Wilson 1987:254–259. Wilson uses a broader definition of hazardous wastes than that used by the British government. He estimates a total figure for hazardous wastes of 2.5 times that of special wastes (Wilson 1987:243).

More recently, Britain has increased its incineration capacity: in 1993 it was reported at 82,000 tonnes per year, with a further 85,000 tonnes under construction or in planning (Yakowitz 1993:142).[62] It already has four specialized high-temperature incinerators for hazardous waste. However, as the OECD reports, much of this new construction will replace older incinerators no longer meeting the necessary requirements (OECD 1994b: 71).

According to the OECD, generation of "special wastes" in Britain has been increasing in recent years, reaching 2.9 million tonnes in 1991/2 (OECD 1994b:65). However, comparing the figures for domestic waste generation to capacity to get an indication of overall capacity is impossible, mainly because waste producers are under no obligation to report the amount of wastes they send out for disposal (Wilson 1987:242). Hence the DoE's figures are based on reports from the WDAs on the amount of wastes they receive on an annual basis. At the same time, very few landfills have weigh-bridges installed that would accurately measure these quantities.

Some observations are possible nonetheless. First, incineration capacity for hazardous waste disposal is extremely overloaded. Using Yakowitz's figures, Britain's incineration capacity is 82,000 tonnes per year, while Britain generates 4.5 million tonnes of hazardous wastes (Yakowitz 1993:142). This observation is backed up by Wilson, who cites the 1985 HWI Report, which stated that there was a severe shortfall in capacity, especially for dealing with the most toxic wastes. This in turn has led to lengthy storage times (with an increased risk factor), and even the landfilling of PCBs in the absence of sufficient incineration capacity (ENDS Report 125, June 1985:11). Landfill methods are also under pressure, as available land dwindles and regulations on its use increase. In addition, the 1980s saw increased EC moves toward blocking the dumping and incineration of wastes at sea. The United Kingdom initially resisted these moves (Yakowitz 1993:141), but has now agreed to implement them. Disposal of industrial wastes at sea ended in 1992; all such disposal was due to be phased out by 1998 (OECD 1994b:72). It should be noted, however, that representatives of the waste industry dispute the claim that British facilities are being used to capacity, asserting that "[hazardous waste] imports have proved vital to the commercial viability of certain (waste) treatment plants, making up for the low utilization by U.K. waste producers."[63]

Assessing Comparative Advantage It would appear from the above that the British waste management industry, with its competitive structure and international outlook, is a natural candidate for engaging in waste importation. It is also the case that Britain does have a comparative advantage in offering waste disposal services. However, contrary to the arguments offered in chapter 3, it is one based on low cost rather than high quality of facilities. The average cost per tonne of hazardous waste disposal in Britain is roughly £50; this varies, of course, according to technology. Landfill costs, for example, are roughly £5–£20 per tonne (OECD 1994b:80).[64] In Germany, on the other hand, landfill costs are more than double that figure (OECD 1993c:54). Furthermore, the figures above indicate that Britain's disposal capacity is very probably overloaded, allowing for little excess "space" for waste importation once domestic needs have been met.

In fact, the highly critical and well-publicized reports issued by the HWI since its inception highlight the fact that Britain's waste disposal industry is fraught with problems, leaving no doubt that policymakers are well-informed about the current situation. Its 1985 report "paints a disturbing picture of falling disposal standards, inadequate licensing of disposal sites, poor enforcement of license conditions, and cut-throat competition which may kill off more of Britain's dwindling specialist disposal facilities" (ENDS Report 125, June 1985:9). It concludes: "All is not well with hazardous waste disposal. Though there is no evidence that hazardous waste disposal is posing unacceptable risk to public health, we are not convinced that the standards and practices widely adopted by the disposal industry provide a sufficient guarantee of protection of the environment" (quoted in Hildyard 1986:220).

Therefore Britain's ability to offer relatively cheap disposal facilities is probably a powerful explanation as to why waste generators decide to export to Britain. However, this factor undermines any claims that Britain is providing specialized, high-quality disposal facilities to the rest of the world, instead supporting the argument made here, that Britain is in fact taking on a considerable burden of risk in taking other countries' unwanted wastes. This in turn would seem to breach the government's duty of care toward the environment and the health of its population. Hence the main problem with this argument is that it fails to explain why the government has continued to support policies of waste importation.

Concluding Analysis and Discussion

To employ a political culture argument very briefly, an observer of the British environmental policy process might well be forced to conclude that it is, above all else, very "British," with all the trappings of "not making a fuss," "gentlemanly" resolution of disputes, and something of a "not in front of the servants" character. Some have pointed out the advantages of the system's flexibility and informality—Moe and Caldwell (1994) and Vogel (1986) argue that Britain has achieved a level of pollution reduction comparable to that of the United States, at a much lower cost and with much less associated bureaucratic red tape. However, these very characteristics have led to practices involving a high degree of (perceived or actual) environmental risk to a population ill-equipped to take on the powers of Whitehall or the big firms in the waste disposal industry. In turn, this yields the conclusion that the British government, contrary to many analyses of unitary states, demonstrates the characteristics of a weak state in the area of environmental policy. Not only is it divorced from the mainstream of popular opinion, but also, through a diffuse regulatory structure and long-standing cooperative relations with industry, it appears to be relatively ineffectual at implementing proposed policy changes. This is apparent not only in the immediate area of waste importation but also, as I discuss below, in bringing more sweeping reforms into effect. At the same time, however, there is evidence that the British government is adapting its strategies to deal with the waste importation problem.

In this chapter I have demonstrated that the importation of hazardous wastes into Great Britain poses a not inconsiderable degree of risk to human health and to local ecosystems. The wastes imported into the country are of a reasonably high level of toxicity; however, existing modes and facilities of waste disposal are not up to the task of adequate disposal. The Hazardous Waste Inspectorate, subsequently part of Her Majesty's Inspectorate of Pollution, has made these points, as well as the absence of proper monitoring capabilities, abundantly clear in its reports.

I also identified a particular configuration of actor preferences regarding the trade. First, the waste disposal industry favors importation, a practice opposed by public opinion and by several vocal environmental groups. At the same time, local authorities and the waste disposal firms under their jurisdiction have little incentive to place controls on waste importation from overseas (and, indeed, a positive incentive to allow them). Central government, on the other hand, which previously supported waste importation, is now trying to implement a ban on these practices.

Thus, in regard to Britain's propensity to import wastes, I have shown that an explanation based on the way certain characteristics of Britain's regulatory system have shaped the behavior of actors and their input into the policy process best captures the dynamics of this situation. Neither contending explanation worked in this case. For example, the economic-nationalist explanation does not capture the diffusion of authority or the costs facing government actors in allowing waste importation to continue. The second explanation, based on comparative advantage in disposal facilities, posited that the legal trade among developed (OECD) countries is determined by level and capacity of disposal facilities: wastes move to countries that have developed a specialization in dealing safely with high-toxicity wastes. The empirical investigation presented here shows that the situation in Britain is exactly the opposite: British firms rely primarily on cheap, low-level disposal facilities, whose capacity, by most calculations, is extremely overloaded.

On the other hand, the British system of environmental regulation conforms to the conditions under which the hypotheses presented in chapter 3 predict would generate high levels of waste importation. Its regulatory structure is highly decentralized, even diffuse. Regulatory powers are devolved to a disparate collection of local authorities, and its waste industry is both privately owned and highly competitive. This diffusion of responsibility has made effective control and coordination of waste importation practices from the center practically impossible. A principal-agent situation exists, enhanced by the conflictual relationship between the central government and local authorities, which scarely have the capacity to monitor wastes generated in their own jurisdictions, let alone those imported from outside. Environmental policy is made within a community that includes government officials and industry representatives but excludes public opinion through a variety of mechanisms, and the highly flexible application of regulations gives industry a good deal of leeway in how it interprets policy. Taken together, these factors are key in explaining the "filtering out" of public opinion and the opinions of environmental groups in related policy decisions, and the extremely tight relationship, which continues to this day, between government officials and the waste disposal industry.

The late-1990s witnessed three changes in government strategy, which if effective should have a big impact on waste management practices in the U.K. First, overall responsibility for hazardous waste management has been centralized. The WRAs have been removed to the supervision of the Environment Agency and stricter inspection policies have been established, steps called for in the much-anticipated National Waste Management Planning Guide and the Management Plan for Exports and Imports of Waste (DoE, 1995, 1996).[65] Plans have also been laid for a National Transfrontier Shipment Service, to be based in Manchester and run by the Environment Agency (ENDS Report 271, August 1997:11). Second, in March 1995 the British Chancellor announced a new tax to be imposed on landfill sites—Britain's first Green Tax—which is expected to raise disposal costs by 50 percent.[66] It is unclear, however, what the final effect will be on the relative use of landfill compared with incineration. Early reports on the effects of the tax showed that the government is likely to reach its revenue targets but also presented evidence that wastes have been diverted to illegal disposal routes. There are few signs that the tax has boosted practices of waste minimization or recycling (ENDS Report 265, February 1997). EU Directives against the use of landfill techniques are also forcing the U.K. government to reconsider its practices (ENDS Report 280, May 1998:21–25).

Finally, in September 1996 new regulations came into force bringing the U.K. definition of special wastes in line with EU regulations (see table 4.2), increasing by almost one-third the quantity of waste treated as special (ENDS Report 255, April 1996:35). These changes have, on the whole, been welcomed by leading firms in the industry, whose position under stricter legislation will be enhanced vis-à-vis the smaller companies. These same firms, however, have been highly critical of the slow pace of institutional reform. Many bodies, as well as the industry, have also criticized the lack of reliable data and statistics on waste generation and disposal, arguing for a national classification scheme (ENDS Report 265, February 1997:21).

If the theory presented here is correct, these changes should lead to an overall reduction in the amount of hazardous wastes imported by British firms, especially via the centralization of hazardous waste regulation. However, there are some counterbalancing factors. First, the pace of change has been remarkably slow: the gap between announcement of policy changes and their final implementation has numbered in years, and progress has been made at an incremental rather than radical pace (Jordan 1993; Aarhus 1995; Voisey and O'Riordan 1997). Moreover, the monitoring capacity of new government agencies remains low.[67] Second, the broader political context remains unchanged. Environmental groups are learning to go to the European level rather than to domestic bodies to further their demands, and the new Labour government pledged to include a wider range of environmental groups into the consultation process (ENDS Daily, February 26, 1997). However, the sorts of institutional changes that would allow greater access to political power at the domestic level for these groups, regardless of which party is in power, are unlikely, despite official promises of "open government," as in Prime Minister John Major's Citizen's Charter. Also, while interest has been expressed in moving toward a more adversarial and/or formalized mode of policy implementation, dominant interests continue to favor principles of "cooperative" regulation (Weale 1996:118). The power of the major importing firms to affect government policy should not be underemphasized, and under these changes their position in the policy process has not been diminished. In chapter 7, I revisit the longer-term prospects for regulatory change as a result of further EU integration, and I consider the prospect of change in international waste trading practices, both within the EU and globally, following the full implementation of the Basel Convention. I also attempt to disentangle the respective influences of globalization, Europeanization, and domestic politics. The current terms of the Basel Convention scarcely affect waste importation to Britain, given, after all, that it is an industrialized country. However, if a full ban on waste exports to LDCs is implemented, then it is likely, in the absence of effective action, that the pressure on Britain's already overburdened facilities from imported wastes will increase.

5
Germany: Technocracy, Federalism, and Risk Aversion

Why, in contrast to Great Britain, is Germany not a net importer of hazardous wastes? In fact, Germany is one of the world's largest exporters of hazardous wastes. Prior to 1990 the bulk of these wastes were exported to what was then the German Democratic Republic (DDR); however, Germany now exports its wastes to other West European countries, primarily France and Great Britain. The German government has also lobbied at the European level against the imposition of a ban on trading wastes for recycling or recovery within the European Union.

This observation generates two puzzles. First, although Britain and Germany exhibit broad similarities—they are both advanced, industrialized democracies with sophisticated and well-established systems of environmental regulation—their observed patterns of behavior with respect to the waste trade differ significantly. Great Britain exhibits highly risk-acceptant behavior, while Germany exhibits highly risk-averse behavior in the face of the same (objectively assessed) degree of international environmental hazard. Second, Germany is known for its strong record on domestic environmental issues and its active role in interstate negotiations over transboundary and global environmental issues. However, its levels of waste exportation run counter to its international reputation as a "Green Knight." A closer examination of the preferences of the different groups of stakeholders in the waste trade reveals further similarities between the two countries, especially among societal groups. As in Britain, there is a high level of societal opposition towards waste importation, and certain elements of the waste disposal industry indicate that they would like to import wastes. Furthermore, both countries have, at first glance, similarly decentralized structures of hazardous waste management and a close relationship between government and industry in the formation of environmental policies.

In this chapter I examine the argument that the extent and nature of Germany's involvement in the waste trade can be traced to its structure and style of environmental regulation, and I test the regulatory account against the comparative advantage explanation. In chapter 4 I showed why the economic nationalist explanation does not apply comparatively. Here I discuss comparative advantage first, as those findings lead into the discussion of Germany's regulatory system.

The strict standards imposed by Germany's environmental regulations and its technocratic administrative structure are reinforced by a high degree of public (pro-environmental) input into the policy process. There are fewer incentives for local authorities to allow waste importation because a more centralized, and federal, regulatory structure reduces the principal-agent dilemmas exhibited in the British case. Therefore Germany does not import wastes, but in fact exports them. These factors work in two ways. First, government-industry and government-society relations are shaped by institutional structures such that industry does not have the same influence on policy decisions as the British waste disposal industry does, and policymaking is more open to different societal groups. Second, costs of waste disposal in Germany are extremely high. This effectively imposes a barrier to trade in such wastes, hence providing a disincentive for countries to export their wastes to Germany. These high costs are primarily a function of the German regulatory system rather than the sorts of technology employed by waste disposal firms.

Dependent Variable: Germany's Involvement in the Waste Trade

Available data from 1986 to the present indicates that Germany has maintained a consistent position as a net exporter of hazardous wastes.[1] In fact, several sources list Germany as one of the world's largest legal exporters of hazardous wastes, both prior to and following unification, as table 5.1 shows (OECD 1993c:190; Pflügner and Götze 1994, tables 2.4 and 2.6). Data from the Umweltbundesamt (UBA) for exports,

Table 5.1
Summary of transfrontier movements of wastes, Germany, 1989–1991

Year	Exports (tonnes)	Imports (tonnes)	Net exports (tonnes)
1989	990,993	45,312	945,681
1990	522,063	62,636	459,427
1991	396,607	141,660	254,947
1992	548,355	76,375	471,980
1993	433,744	78,219	355,525

Source: OECD 1997a, table 1. The sharp decrease shown from 1989 to 1990 is due largely to German unification in 1990. Only wastes going to final disposal have to be reported.

summarized in table 5.2, are comparable. UBA figures for 1994 show that Germany exported 336,445 tonnes of special waste under the EU waste shipment regulations and the Basel Convention and imported 71,080 tonnes (*Haznews* 105, December 1996:12). Table 5.3 shows 1996 net exports of 568,165 tonnes of hazardous wastes. As was the case in Great Britain, waste accounting in Germany had shifted to using Basel Convention lists by 1996.

In 1990 the OECD lists the total amount of hazardous waste managed in Germany in relation to generation as 92.3 percent; in 1991 the figure is 95.8 percent (OECD 1994a:table 4). The OECD notes that its figures for waste exports, based on those reported by the German government, are extremely conservative, as they "do not include exports of certain types of residual matter which are considered as hazardous under the legislation of the importing country or under the OECD Decision on transfrontier movements of hazardous waste [C(88)90]. On the basis of export data from countries such as Austria, it may be assumed that residual matter exported from Germany *probably amounts to a quantity equal to the quantity reported as hazardous wastes in German statistics*" (OECD 1993c:191; emphasis added).

Prior to 1990 most German wastes were shipped to the DDR.[2] In addition, Germany exports to France, Britain, Switzerland, Austria, Belgium, and the Netherlands. Exports to Britain have increased tenfold since 1990.[3] The German Länder also differ significantly in terms of their exporting practices. The lead exporter of wastes abroad is Baden-

Table 5.2
German special waste exports, 1988–1993

	1988	1989	1990	1991	1992	1993
Belgium	128.3	81.5	158.6	153.0	214.1	164.5
Denmark	—	0.4	11.3	7.6	2.7	28.4
Finland	—	—	—	0.2	0.3	0.5
France	197.5	181.7	154.1	223.4	112.1	124.4
Great Britain	36.2	2.9	0.6	3.3	9.5	55.5
Luxembourg	—	—	—	—	19.0	10.8
Netherlands	11.4	71.8	176.3	72.0	283.7	221.0
Norway	—	—	—	28.0	—	—
Poland	—	—	—	—	—	1.1
Switzerland	2.7	0.2	2.0	0.1	0.9	5.7
Total	376.1	338.5	502.9	487.4	642.3	611.9

Source: UBA, Fachgebiet III 4.1. Figures in 1,000 tonnes. Data from 1988 to 1990 only for the old Bundesländer, and do not include exports from the Federal Republic to the DDR. Data from 1991 includes figures from the new Länder. Figures calculated on the basis of the Waste Disposal Law valid until 1993.

Table 5.3
Transfrontier movements of waste, Federal Republic of Germany, 1996

In tonnes	Exports	Imports	Net exports
For recycling	1,107,895	253,564	854,331
For disposal	112,183	93,570	18,613
Total	1,220,078	347,134	872,944
Subtotals:			
Municipal waste	28,943	1,065	27,878
Hazardous wastes[1]	821,718	253,553	568,165

Source: UBA, *Umweltdaten Deutschland 1998*, p. 43.
Note:
1. Under Annex I of the Basel Convention.

Württemburg, followed by Hamburg (Pflügner and Götze 1994; Vallette and Spalding 1990). Germany also has a significant problem with illegal exports of hazardous wastes: Greenpeace and other international organizations have catalogued many occurrences in recent years of illegal transportation of wastes across national frontiers.[4]

The case of German waste exportation to France is emblematic of the ways in which problems with waste management at the domestic level have spilled over into the international arena. For many years, this practice had been endorsed by both sides as a way for the German states bordering onto France to (temporarily) bypass capacity shortages and high disposal costs. Building new incineration capacity, in particular, was hampered by the activity of environmentalist groups. However, in 1992 it was discovered that a shipment of wastes from Baden-Württemberg to France contained highly toxic substances not listed on the accompanying documentation. This provoked the French authorities to impose a ban on waste imports from Germany, an action that had, as I discuss below, an important impact on the development of strict waste minimization policies in the Federal Republic.

Comparative Advantage in Waste Disposal?

Germany has a very clear comparative advantage in waste disposal. It offers high-quality, state-of-the-art facilities to treat even the most toxic wastes. Recent evidence also seems to suggest that fairly significant spare capacity exists, at least in certain Länder, although this assertion is strongly contested by some sources. According to this explanation, Germany should therefore be more likely to import hazardous wastes than countries with lower quality facilities (such as Great Britain). However, this argument does not hold in the German case.

Disposal Technology

Germany is notable for relying more on incineration than on landfill techniques for the disposal of hazardous wastes, compared with its European counterparts.[5] According to an early estimate, "it is estimated that three to four million tonnes of 'special wastes' per year are being generated in the FRG. About 15 per cent are disposed by incineration,

35 per cent treated by chemophysical treatment and 50 per cent disposed of in secure landfill sites" (Defregger 1983:16).[6] 1990 figures from the Umweltbundesamt (UBA) indicate that there were 14 landfills open for public use, 29 major facilities for the incineration of special waste, and roughly 70 physicochemical treatment plants.[7] Many of the major firms—BASF, Bayer, and Hoechst, for example—have developed their own highly sophisticated systems for onsite waste disposal, while, as mentioned above, several Länder operate state-of-the-art integrated treatment facilities for local firms.[8] Landfill requirements in the FRG are stringent, unlike the situation in the U.K. Sites must be either lined artificially or constructed in areas where the soil is impermeable (Sierig 1987:129; OECD 1993c:56), and the practice of codisposal is considered unacceptable according to public health and safety requirements.[9]

Disposal Capacity

The disposal capacity issue in Germany is an interesting one. From the 1980s until the early-1990s, it was generally considered that the country was facing an overall disposal capacity crisis. Now, however—for a variety of reasons I discuss later—this appears no longer to be the case.

Landfill availability is definitely in decline: the UBA estimates that currently available landfill space will be exhausted in the next few years. Hence government authorities are currently seeking to expand total incineration capacity: "Present hazardous waste incineration capacities amount to about 1 million tonnes per annum, estimates of capacities required in the year 2000 are set at 2 million tonnes p.a." (Schmitt-Tegge 1994:443). The Federal Environment Agency expects that to meet this desired increase in capacity, roughly 13 new facilities need to be constructed, at an expected cost of DM 5 billion in the next eight to ten years (ibid.:446). Increased underground storage capacity is also being planned (Czech 1996). These moves, however, are being met with a high degree of popular opposition (Schmitt-Tegge 1994:441), and it is by no means certain that the German government will be able to achieve a satisfactory solution to the NIMBY dilemma.

At the same time, many reports indicate that due to a decline in the amounts of hazardous wastes being produced by German firms, a fairly high proportion of German disposal capacity is underutilized. Linnerooth and Davis discuss the implications of a declining waste market for disposal facilities in Hessen (Linnerooth and Davis 1987:183–184), and OECD figures indicate that waste quantities have not increased since 1980 (OECD 1993c:53). More recently, figures released in January 1996 stated that "German waste sites and incinerators operated at only 65% of capacity in 1995."[10] The 1994 Annual Report of the Federal Environment Agency estimated that waste generation will decrease by up to 50 percent in the years to come (Czech 1996). In general, however, the perception is that there is a lack of space in Germany to dispose of wastes, heightened by the dense concentration of population and industrial activity in the country as a whole.[11]

Assessing Germany's Comparative Advantage: The Role of Costs
An analysis of Germany's waste trade practices based on the comparative advantage thesis would be forced to conclude that the reason Germany exports rather than imports wastes is that the costs of waste disposal are prohibitively high. For example, according to Defregger, "In the FRG, the charges for hazardous wastes that can be landfilled range from 50 to 100 German Marks per tonne, while the price per tonne for particularly hazardous wastes, requiring chemophysical or thermal treatment, may reach several hundred German Marks" (Defregger 1983:21). This trend is borne out by recent data from the OECD. While "at the beginning of 1990, western German landfill prices for metal sludges, oily sludges and asbestos were roughly double the median price for the rest of Europe; incineration prices ranged between DM450 to DM2000 per tonne for hazardous wastes, and over DM300 per tonne for physico-chemical treatment of several special wastes" (OECD 1993c:55).[12]

While this explanation works to a certain extent, several questions remain unanswered. First, why are disposal costs so high in Germany? Comparisons with other countries show that disposal costs in Germany for the same sorts of wastes, and for the same sorts of disposal techniques, are considerably higher than in equivalent countries. One study puts the average costs of landfill in Germany as ranging between 113–425 ECUs per tonne for dangerous wastes; in France the average cost is 11 ECUs per tonne. For incineration, German costs ranges between 62–

348 ECUs per tonne; in France the average is 34–39 ECUs per tonne (Pflügner and Götze 1994:110).[13] This suggests that something other than the logic of the free market is intervening here. Second, why isn't Germany taking on the wastes generated by countries that cannot dispose of those wastes as safely? Germany's strong record on the international environmental scene (see below) would suggest that this would at least be considered as a policy option. However, there is no evidence of such deliberation taking place. Instead, this risk-averse behavior is best explained by Germany's system of environmental regulation. In the following sections of this chapter, I answer some of the questions raised by the above account and provide a more complete explanation of why Germany is not a net importer of hazardous wastes.

German System of Environmental Regulation

Historical and Institutional Context

The history of environmental awareness and subsequent political action in Germany over the past fifty years has been heavily conditioned by two broad factors. First, the extraordinary reconstruction and growth of the German economy—the *Wirtschaftswunder*, or economic miracle—brought with it a heavy burden of environmental degradation, including the heavy pollution of Germany's main rivers, severe land degradation, and air pollution. Perhaps the most critical symbol of the costs of economic growth was the damage to, and indeed death of, a large part of its forest cover (*Waldsterben*) through acid rain deposits. For many Germans, this crisis (which continues today) was one of the key focal points during the 1980s for renewed environmental awareness, both at the national and global level.

Second, Germany has always been particularly vulnerable to pollutants, especially air pollutants, originating outside of its borders, and has therefore often played a strong role in international environmental negotiations and in the formation of EC/EU-wide environmental policy. Yet, as observers point out, until the late 1970s Germany was something of a laggard state in global environmental negotiations. Only when the damage done at home by pollutants originating from overseas become distressingly apparent did the government begin to play a more proactive role in setting and implementing related targets.[14] Germany also dominates the international market for environmental technology, having become a major supplier of equipment for the reduction of pollutants at source (OECD 1996). As a result, Germany has earned a reputation as the "Green Knight" of Europe (Rüdig 1993), noted for some of the most stringent domestic environmental regulations within the OECD, as well as for its activist approach on the international scene. In the following sections of this chapter, I show how Germany's regulatory structure and style have determined its involvement in the waste trade, and I address why its "green" reputation does not extend to this area.

Structure of Environmental Regulation in Germany

General Overview of Regulatory Structure The structure of environmental regulation in Germany must be understood in the context of the country's federal structure. The Federal Republic consists of sixteen Länder: eleven "original" states and the five new Eastern states. The Länder are represented at the national level in the Upper House of Parliament—the Bundesrat—which has supervisory and veto powers over most legislation proposed by the Lower House—the Bundestag. More fundamentally, in terms of public policy, powers of policy formation and implementation are split: "With few exceptions, such as defense and foreign policy, public administration is not a responsibility of central government but of the Länder, local government and parapublic institutions" (Schmidt 1996:81). Characterizing German public policymaking as "intergovernmental," Schmidt continues: "Public policy in the Federal Republic rests upon an exceptionally high degree of intertwining of policy-making between the federal authorities and the Länder, comprising assured participation rights and the veto powers of the Bundesrat in major legislation, and joint planning and demonstration between the federal government and the state governments on a wide variety of matters that require cooperation between the federal government and state governments, and for which they have shared responsibility, as in education, financial planning, economic development, science and research, and environmental policy" (83).

In terms of states' fiscal and other resources, the shared responsibilities between Bund and Land levels of government are matched by shared tax revenues. Measures exist to ensure that no Land, or local authority within a Land, receives more than its fair share of overall revenues. Furthermore, the Länder have an additional source of authority in that they supply most public services and employ the majority of Germany's public servants.[15]

One result of Germany's constitutionally protected division of powers is that, unlike in Britain, "the freedom of action of the federal government is heavily circumscribed" (Schmidt 1996:90). Partly as a result of this, intergovernmental relations have been characterized by a much higher degree of cooperation and stability in Germany than in Britain. Each level of government is able to act within well-known and accepted constraints, and from a position of fiscal and electoral security, although certain strains have emerged in this relationship since 1990 due to the relative poverty of the new Länder in comparison with their western counterparts.

The earliest environmental initiatives were developed in the nineteenth century prior to unification of the individual states (now Länder).[16] Owing to this decentralization of political authority, transboundary environmental issues—especially river pollution—soon became an obvious problem. In response, the states established loose coalitions of businesses and regulatory authorities across Land borders to regulate emissions into individual river systems; these *Genossenschaften* were probably among the earliest examples of transboundary environmental agreements, and they remain in place to this day (Weale 1992b:162).

This pattern of state-level initiatives, often copied across the Länder, continued throughout the turbulent first half of the twentieth century up until the early 1950s—the years of the *Wirtschaftswunder*, when economic goals, in particular that of reconstruction, were given priority over environmental goals. It was not until this time that the federal government consented to play a role in determining and enforcing national standards of environmental protection.

Abromeit (1990) divides the post war progress of environmental legislation into three periods. The first (1949–69) was characterized by industry playing a lead role in extending environmental regulation. Key legislation at the federal level during this period included the Federal Water Resources Act of 1957 and the Clean Air Maintenance Law of 1959. Both of these followed the same pattern, whereby users had to obtain permits in order to operate and set acceptable emissions limits (Weale 1992b:163). The second period (1974–82) was marked by the growing politicization of the environment following the 1972 Stockholm Conference, severe pollution of the Rhine as a result of toxic waste dumping, and the beginnings of the German environmental movement. During this period, the federal government began to play a more active role in environmental regulation, including establishing in 1974 the Umweltbundesamt (UBA), or Federal Environment Agency, which, unlike the U.S. EPA, is not a federal ministry, but rather "purely concerned with research and advice" (Weale 1992b:165).

The third period identified by Abromeit began with the 1982 election of the pro-industry CDU/CSU-FDP coalition headed by Helmut Kohl and was characterized by closer cooperation between government and industry in the environmental arena. The most important legislative development in the 1980s was the establishment of a national ministry for the environment—the Bundesministerium für Umwelt, Naturschutz und Reaktorsicherheit (BMU)—in 1986, largely as a response to the outcry following the Chernobyl disaster. The BMU took over the various environmental duties that had hitherto been scattered among different government ministries.

In fact, environmental issues are emblematic of the sorts of cooperative intergovernmental relations that are a central feature of contemporary German politics. The Grundgesetz (federal constitution) does not specify how powers of environmental protection should be divided among Bund and Länder. Hence competencies vary across different environmental issues, according to whether the federal government has more or less control over policy formulation. Länder are always responsible for implementation; hence, the Bundesrat has powers of veto and amendment over environmental legislation. According to the OECD, "in most cases, federal environmental law supersedes Länder laws. Areas which are overwhelmingly regulated by federal law include nuclear power, waste management, air quality management and noise abatement. Nature conservation, landscape protection and water management, on the other hand, are areas where only framework legislation can be passed at the federal level. The more specific regulation of these areas remains in the domain of the Länder" (OECD 1993c:26).[17]

At the Land level environmental management is under the control of the Land Environment Ministries (Bothe 1986), which oversee activities at the regional and local level, usually according to medium (air, soil, water). There is little variance in this basic structure across the Länder, with the exception of the eastern states, which are still adapting to western patterns. Finally, several consultative bodies—notably, the Conference of Environment Ministers—serve the need for coordination of activities and practices among the Länder, which consists of Land and federal officials. As Héritier et al. argue, "Partly through this federal state structure, which contains elements of cooperation and competition, there are a number of political levels in the Federal Republic which, while they operate fully independently of each other, are also integrated in a number of ways. In the area of air pollution [as with waste management], both vertical and horizontal forms of political integration matter: for example at the inter-governmental level (the Conference of Environment Ministers), the parliamentary level (the Environment Cabinet), and the administrative and the party political levels."[18]

Structure of Hazardous Waste Management Hazardous waste management in the Federal Republic is primarily the responsibility of the Länder (Linnerooth and Davis 1987:154). The German model, such as one exists, is perhaps most accurately characterized as mixed private-public activity, with no clear separation between the roles of private firms—both waste generators and disposal firms—and public bodies. Management practices vary extensively from Land to Land. Capacities for waste disposal, in terms of available resources and personnel, also vary from Land to Land. The clearest differences are between the older, western Länder and the new Länder, formerly East Germany, which not only face extensive cleanup issues but also require large injections of resources to bring their industrial and environmental base up to the same level as the western part of the country.

In terms of overall strategies and legislation, the Bund plays an active role beyond the formulation of framework legislation, including imposing a duty of care on Länder to ensure that waste management reaches certain nationally acceptable standards. As in Britain, hazardous waste regulation has been in place in Germany at the federal level since 1972, when the *Abfallbeseitigungsgesetz* (AbfG), or Waste Disposal Law, was first implemented (Czech 1996). Although some legislation at the Land level had been in place earlier, the AbfG meant that for the first time uniform and binding provisions with respect to waste disposal were brought to bear on the Länder.[19] The law regarding hazardous wastes has subsequently been amended five times (Czech 1996). In particular, the third amendment of 1985 introduced permits for the export and transportation of wastes, and the 5th Amendment of 1994 (to be promulgated in October 1996) contains important measures regarding the avoidance and recycling of wastes, introducing the concept of the *Kreislaufwirtschaft*, or "closed circle economy."[20]

The Länder are therefore responsible for devising waste management plans in order to implement national legislation and to meet centrally imposed standards. Municipal waste management is the province of lower levels of government.[21] The common elements of these plans are lists of the kinds and quantities of waste to be treated within the area, the number and types of disposal facilities in operation, and the management of these facilities (Defregger 1983:16). Beyond this common framework, actual practices vary extensively.

In terms of managing hazardous waste disposal, some Länder operate publicly owned or publicly supported, integrated facilities, while others rely on a larger number of privately owned, decentralized facilities (Linnerooth and Davis 1987:156).[22] In some cases, waste generators also have an ownership stake in separate waste disposal companies. The two main examples of states with publicly run, integrated facilities are Bavaria—the largest Land in the federation—and Hessen. As waste producers in these two states are obliged to send their wastes to local facilities, the corporations that run those facilities—in Bavaria, the Mittelfranken Cooperative for Special Waste Management (ZVSMM) and the Management of Special Wastes in Bavaria (GSB), and in Hessen, the Hessian Industriemüll GmbH (HIM)—hold a practical monopoly on hazardous waste disposal corresponding to the public monopoly model developed by Linnerooth and Davis (1987). In North Rhine-

Westfalia, a heavily industrialized region encompassing part of the Ruhr industrial belt, and in Lower Saxony, small, privately held firms provide most waste disposal services, although some publicly run facilities exist.[23] Information on other Länder is more scarce.[24] However, generally speaking, mixed models are the norm, as are high levels of government funding or subsidy (OECD 1993c:55).[25]

In sum, the structure of hazardous waste management in the Federal Republic of Germany can be characterized as a decentralized but coordinate model involving a combination of levels of government and a mixture of private and public enterprise. However, the German model of hazardous waste management (including both disposal and regulation) is nowhere near as diffuse as the British model. Several factors attest to the much higher level of coordination in the German structure of hazardous waste regulation.

First, the Länder are subjected to a much stronger level of supervision from central authorities than the British local authorities in terms of statutory requirements they must show evidence of meeting.[26] The cooperative relationship between the Länder and the Bund, too, has meant that national-local preferences have not diverged as significantly as they have in the U.K. Also, the federal government has not sought to undermine the power of state governments in any way (nor can it, given the constitutionally imposed restraints on its freedom of action). In turn, this relationship has also meant that problems of uncertainty and asymmetric information witnessed in Great Britain are nowhere near as severe in the German case.

Second, there are much higher levels of government intervention in the waste industry in Germany than in Great Britain. For example, the industry itself is run much more on a public service than a competitive basis, and in several (although not all) Länder, industry structure is much closer to monopoly. A long tradition of public ownership or management of disposal facilities, using a mix of different regulatory strategies such as subsidies and policies aimed at restricting and/or controlling the movement of wastes among the Länder, has created a very different operating environment for German waste disposal firms.

Third, the capacities and abilities of the Land governments to regulate and control the importation of wastes are, on the whole, high. Local authorities are in a strong position in Germany: they have at their disposal adequate revenue to avoid the need to turn to outside sources of funds and to ensure that waste disposal sites are correctly managed and do not violate existing laws and regulations. Germany has only 50 waste regulatory authorities in the entire country (Pflügner and Götze 1994), compared with close to 200 in the U.K.; thus it has much less of a problem coordinating the monitoring and control of movements of waste.

Germany's Style of Environmental Regulation
Kitschelt (1986) characterizes Germany as having a closed political opportunity structure.[27] On first glance, the procedures for actual policy formation appear relatively closed: details are worked out between government actors, technical/scientific advisors, and industry representatives. However, my analysis here shows that in fact a variety of factors combine to make the policy process in Germany more responsive to a broader range of participants and interests than that in Great Britain. Formal mechanisms enable a wide range of groups to be incorporated into the policy community, and Germany's electoral system, its federal division of powers, and its independent judiciary afford multiple points of access for a broad range of societal groups. Furthermore, its mode of policy implementation demonstrates a high degree of risk aversity and embodies the German tradition of state intervention to control the extent to which industry is able to engage in practices that degrade the environment.

Process of Policy Formation: A Narrow Policy Community In Germany, environmental policies and detailed prescriptions and goals are worked out within a relatively narrow policy community. Several analysts (e.g., Kitschelt 1986, Weale 1992b) argue that the political opportunity structure for public participation in environmental policymaking (the input side of the policy process) is relatively closed. They argue that two main groups of nonstate actors are involved in working out detailed prescriptions and goals for environmental policy: industry actors and scientific/technical experts.[28] German "environmental policy-making has been based on two main pillars: the tradition of cooperation between government and industry and the participation of experts and scientists in the formation of policy" (Aguilar 1993:234), an arrangement often labeled as corporatist.[29]

Industry access to policy rests on the "cooperation principle" (*das Kooperationsprinzip*) and voluntary agreements in policy negotiations between government actors and industry organizations. The cooperation principle, according to the BMU, provides a means whereby the government can involve societal groups in the conception and implementation of the goals and instruments of environmental policy (Héritier et al. 1994:61). However, in practice industry groups—due to their greater resource mobilization capacities—have been able to dominate other groups—especially environmental groups—in this process. For example, industry groups have more access to and say in the formation of working groups on environmental policy, as well as tighter informal connections with government officials (ibid.:61–4).[30]

Government and industry groups work together in policy formulation through a process of voluntary agreements, whereby the government publicly announces a set of policy objectives that firms (represented by peak associations) can choose to either implement voluntarily or have imposed upon them. Government officials then enter into discussions with the firms who have agreed to the objectives; "these rounds generally end up in industrial compromise (written or oral, but always nonbinding) for the realization of environmental goals. Compromises are later given publicity so that public opinion can somehow act as their guarantor" (Aguilar 1993:235). So far, such agreements have occurred in the regulation of detergents and of other chemicals—notably CFCs (Rose-Ackerman 1995:68–69).

The hard sciences are also well represented in the policymaking arena: Weale speaks of a "cult of expertise" in German politics (Weale 1992b:178). Panels of expert and scientific advisors are given a prominent role in regulatory activity, resulting in a very technocratic approach to policy implementation (see below). This procedure has been accused by some of lacking in public accountability, and hence legitimacy, as well as emphasizing the scientific and technological aspects of environmental problems over the socioeconomic and political aspects (Rose-Ackerman 1995:65, 71; Hager 1995).[31] Of the groups of experts contributing to the policy process, two—the Council of Economic Experts and the UBA—are officially appointed and funded. In addition to these, there exist literally hundreds of private groups that are called upon from time to time for consultation in different issue areas.[32]

The incorporation of many and various expert scientific groups into the policy process at a formal level in Germany to an extent outweighs the possibility that industrial groups might be able to undermine the stringency of German environmental policies. The role of scientific advisory groups is much more important in Germany than in Britain (Boehmer-Christiansen and Skea 1991:48); this in turn has meant that the Germans rely on higher levels of formal modeling and prediction than their British counterparts.[33] This process affords increased legitimacy to German environmental policies and goals, and it also implies that overall assessments of environmental risks are likely to be higher than in a case where industry representatives play a more prominent role.

However, Héritier et al. point out several ways in which this close-knit community is opening up (Héritier et al. 1994:65; see also Blühdorn, Krause, and Scharf 1995). A new generation of public servants who have strong affinities with the environmental movement and are more likely to include such groups in the policymaking process is coming to the fore. In other words, the system is not inherently biased toward the exclusion of environmental groups. Héritier et al. explicitly compare the "winner-takes-all" (or zero-sum) situation prevailing in Britain with the more open German system. In the latter, multiple points of access at different political levels usually gives most groups the opportunity to air their views and have them recognized in policy outcomes (ibid.:65).

Broader Political Opportunity Structure The institutional makeup of the German political system affords to interested parties several points of access to the environmental policy process. First, Germany's electoral system is based on principles of proportional representation. As with other such systems, it is not truly proportional (in terms of parliamentary representation based on the percentage achieved by each group in the national popular vote), but rather a mixed system. German citizens have two votes: the first is cast for constituency-based candidates, the second for party lists in each Land. Half of the 656 seats in the Bundestag and the Land parliaments are allocated according to the results of the constituency votes, the other half through party lists in each of the Länder (Schmidt 1996:66–67). The main restriction on proportionality is the five percent threshold: parties can only win national representation if their share of the national vote is five percent or more, or if they win three or more constituency seats. This has the effect of excluding small (and often extremist) groups, but, as the experience of the Greens has shown, does not provide insuperable barriers to entry at the national level. As a result, Germany tends to be governed by coalition: usually either the CDU-CSU or the SPD in coalition with the FDP, a much smaller party that often plays the "kingmaker" role in postelection bargaining.[34] In turn, the high level of interparty competition in Germany, as well as a heightened awareness of the need for coalitions, has made the larger parties more responsive to the views and strengths of smaller parties than they would be in a less proportional system (Boehmer-Christiansen and Skea 1991:113). Most significant is the opportunity the electoral system afforded to the Green Party (die Grünen) to obtain representation at the national and local level. Their roller coaster ride in terms of electoral fortunes (which I discuss later) played a significant role in altering the political climate in this area, not only through direct policy effects but also by altering the policy platforms of the other main political parties.

Second, Germany's federal system of government, as well as being important in understanding its regulatory structure, also affords further opportunities for access to the policy process in Germany. In federal systems it is generally easier for local groups to work at the state level to get issues onto the political agenda. The nature of the German system, given the importance of the Bundesrat in policy formation and the Länder in policy implementation, is that local issues can then find their way onto the national political agenda (Boehmer-Christiansen and Skea 1991:96). Thus "regional problems in particular have a better chance of succeeding on the political agenda, and of being dealt with closer to those affected than in countries such as Great Britain, with unitary systems, where local authorities are strongly dependent on central government in their range of action" (Héritier et al. 1994:52).[35]

The final point of access to the environmental policy arena is through the German system of judicial review. As in the U.S., the judiciary forms a separate arm of the government, although its role and practices differ.[36] The highest legal authority in the Federal Republic is the Federal Constitutional Court (*das Bundesverfassungsgericht*), which exists to protect the basic legal and political framework of the 1949 Constitution. The Court cannot act of its own accord; rather, there are several main routes whereby policy decisions become subject to judicial review, most commonly complaints registered by individuals who claim violation of a protected right by a public authority (Schmidt 1996:86).

The court system played an important role in setting environmental regulations in the 1970s. The lower-level administrative courts in particular adjudicate facility permitting, although their powers of jurisdiction have been somewhat reduced in recent years (Jänicke and Weidner, 1997:144). In general, the German judiciary upholds and interprets the rather abstract principles underlying German environmental policy, such as the "precautionary principle" (see below), as well as providing a further point of access for societal disputation of government policy. Thus the existence of an independent judiciary contributes to Kitschelt's assessment of Germany's political output structure as relatively weak. Policy implementation by the Länder (contrary to popular opinion) is open to societal disputation at a number of levels via the courts and through the decentralized (federal) system of implementation.

In sum, the regulatory style in Germany is relatively open—an analysis that runs counter to Kitschelt's assessment of Germany's political opportunity system as closed. When broader institutional opportunities are taken into account, it is clear that the German system is more open than Britain's to a wide range of societal interests. While the state to an extent can decide whom it consults in the immediate process of policy formation, constitutional measures also constrain it from systematically excluding certain elements. In addition, the German system in recent years has exhibited certain elements of dynamism (or flexibility) that contribute to the argument that it is not inherently biased against the incorporation of societal interests. One of these was the success of the Green Party, which opened doors to other environmental groups and made the ruling coalition aware of the need to take pro-environmentalist opinions into account. The Green Party's subsequent fall from grace paradoxically strengthened the hand of some environmentalists, as environmental strategies switched away from public protest toward more "professional" lobbying tactics carried out by the largest environmental NGOs. This analysis therefore takes issue with Kitschelt's assessment that the antinuclear protest movement in Germany (from which the Green Party arose) had a greater structural than procedural effect. Rather, I have demonstrated in this chapter that the opposite occurred: political structures remained constant while opportunities for participation broadened.

Policy Implementation: The Rechtsstaat and the Precautionary Principle
Especially in contrast to its British equivalent, the German mode of policy implementation is best characterized as rigid. It is also the area of German environmental policy where the underlying philosophy of German environmental regulation becomes most apparent.[38] The Germans employ a system of legally imposed environmental standards—uniform across industries and formulated through a process of negotiation with industry—"defined by a strong emphasis upon constitutional formalism, on the one hand, and technical expertise on the other" (Weale 1992b:179).

This practice follows a general preference for constitutionally and legally grounded guiding principles for policymaking. Much is made of the notion of the *Rechtsstaat*: "the requirements that state action be conducted in a publicly justifiable manner through legal processes, to which the state itself is subject" (Weale 1992b:177; see also Dyson 1992). To that end, there exist two main guiding principles behind the administration of environmental policy in Germany: the precautionary principle (*das Vorsorgeprinzip*) and the polluter-pays principle (*das Verursacherprinzip*).

The precautionary principle, although not restricted to German policy practices, is manifested most strongly in the Federal Republic and underlies German risk management practices with respect to environmental degradation. It mandates that action to protect the environment be taken in advance of actual damage—"that environmental degradation is to be avoided in the first place, and that environmentally-friendly practices already in place at the levels of construction and production are preferred" (Héritier et al. 1994:29).[39]

The precautionary principle first arose in German policy documents in 1976, and its exact interpretation in both the theory and practice of German environmental regulation has been the subject of much analysis.[40] Jordan and O'Riordan argue that in practice risk-avoidance measures are tempered with cost and technology constraints, and that the principle is implemented through a requirement on firms to use the "very best technology" (*Stand der Technik*). They go on: "For the Germans then, precaution is an interventionist measure, a justification for the state's involvement in the day to day lives of its citizenry in the name of wise environmental practice" (Jordan and O'Riordan 1995:68).[41]

The polluter-pays principle, too, is significant in German practice. It denotes not only the imposition of financial penalties on polluters, but also that the burden of responsibility rests in general terms with those who generate environmental damage (*die Verursacher*).[42] In the arena of waste policy, its implementation can be seen in the 1990 Packaging Ordinance, whereby firms are obliged to take back for recycling or reuse any packaging materials used to ship their products. This application of the principle has now been extended to the hazardous waste sphere with the enactment of the *Kreislaufwirtschaft* ordinance. Firms that generate hazardous wastes are now unambiguously responsible for ensuring that they are correctly reused, recycled, or disposed of.

Thus, in terms of the politics of hazardous waste generation and disposal, the German preference for strict standards and for proactivism in averting environmental damage before it occurs have shaped the policy debate in fundamentally different ways than the debate in the U.K. First, the quality standards imposed on waste disposal firms (for example, the strict pretreatment, landfill, and incineration requirements) are key in understanding why disposal costs are so high in the FRG. Second, the regulatory focus has shifted in recent years toward upstream regulation of wastes. In other words, waste generators are being encouraged to reduce the amount of wastes they generate, either via direct waste reduction or through recycling waste materials back into the production process (OECD 1993c:64–66). As demonstrated earlier, this policy is meeting with a high degree of success, a function of both the superior environmental technology sector in Germany and the high degree of industry's acquiescence with the basic principles underlying German modes of environmental regulation.

In their crossnational comparison of chemicals regulation, Brickman, Jasanoff, and Ilgen provide a good summation of the different modes of environmental policy implementation in Britain and Germany:

> Britain and Germany differ strikingly in their preference for law and formality in defining regulatory relationships. The British carry flexibility to the extreme, developing policy wherever possible through close, informal contacts among government officials and private groups. Flexibility characterizes policy outcomes as well, with guidelines, recommendations and informal persuasion substituting as far as possible for statutory orders and prosecution. The Germans, by contrast, insist on a precise formulation of public and private regulatory responsibilities, usually by means of statutory instruments. The national bias towards precision is reflected both in the consultation process, which tends to be more structured and formal than in Britain, and in the use of numerically exact standards for chemicals in the workplace and pesticide residues (Brickman, Jasanoff, and Ilgen, 1985:53).[43]

In practice, then, the German mode of policy implementation has led to a strongly technocratic approach to pollution control, where legal and scientific advisory bodies play an important role in the development and implementation of standards. The reliance on generalized regulatory principles underlies the extremely conservative approach taken by German regulatory authorities in the calculation of what constitutes acceptable environmental risk.

Germany as Nonimporter of Hazardous Wastes: The Institutional Perspective

Germany, unlike Britain, is not a net importer of hazardous wastes, and indeed exhibits a strong propensity to export its hazardous wastes. In the first sections of this chapter, I argued that an explanation of this phenomenon based on Germany's apparent comparative advantage in waste disposal does not adequately capture the dynamics of the situation. Instead, an institutional explanation that focuses on the nexus of government-industry-society relations offers a better account of Germany's waste trading practices.

Due to some apparent similarities between Germany and Great Britain, this explanation might appear problematic. First, it seems on first glance that Germany's structure of environmental regulation is somewhat similar to Britain's, in that the powers for controlling transfrontier movements of hazardous wastes, as well as those for ensuring the adequate handling and disposal of such wastes, are delegated to subnational authorities. The second apparent similarity is that government-industry relations in the Federal Republic are, as in the U.K., very close: some even claim that a state of "regulatory capture" of government agencies by industrial interests exists in the environmental sphere (Abromeit 1990:65). This closeness—reflected in the ways that industrial interests are incorporated into the process of policy formation, as outlined above—is one of the main focal points of action by Germany's environmental community, who argue that societal voices are excluded from this process. Indeed, the characterization of German environmental policymaking as technocratic rather than participatory is justified in many ways (Hager 1995).

In the next few sections I address these issues, arguing that certain key features of Germany's system of environmental regulation, reinforced by the broader institutional structures in which regulatory frameworks are nested, have shaped government-industry-society relations in ways that have inhibited waste importation: precisely the opposite of the British case. These differences are even more striking in the face of similar preference configurations on the part of key actors, and, indeed, they extend to very different patterns of policy in the hazardous waste management sphere in general.

Die Grünen and Public Opinion: The Politics of Access

Despite the fact that the actual policy community in the Federal Republic has historically been relatively narrow, public opinion and the preferences of environmental groups have, through various institutional mechanisms, been able to make their presence felt in the mainstream political arenas. Such access points have been crucial in shaping the content and scope of German environmental policy, in the same way that their absence in Great Britain has shaped policy there.

The environmental community in the Federal Republic is extraordinarily diverse, encompassing a broad range of different groups and organizations operating in a political culture receptive to environmental opinion.[44] Environmental groups range from locally based grassroots activists to extremely well-funded, highly trained expert groups to the Green Party (die Grünen). It is widely considered that the German public is one of the most informed with respect to environmental concerns. A Eurobarometer study shows that the percentage of German people regarding the environment as an urgent concern rose from 84 percent in 1988 to 88 percent in 1992 (Peattie and Ringler 1994:217).[45]

As in other countries, public concern about hazardous waste ranks highly among environmental issues. One study (cited in Heidenheimer, Heclo, and Adams 1990) shows that on a concern index of 1 to 3, worries about disposal of industrial and chemical wastes score 2.25.[46] A 1991 study of residents of Baden-Württemberg reveals that public attitudes toward hazardous wastes are confused and fearful: "people don't know much about it, but they know that it's dramatic." This study also found that "70 per cent of those questioned were against the export of hazardous wastes, but that 60 per cent said they didn't want it dumped or burned in their own country."[47]

One result of this contradiction has been that public opposition to the construction of hazardous waste disposal facilities has translated into an effective citizens' movement, which in turn has led to increased exportation of hazardous wastes as a "quick fix" to the problem: "In 1992, reports on waste exports, a shortage of waste management capacities and illegal transport of wastes abounded in the German press. On the other hand, citizen action groups and environmentalists claimed as their big achievement to have prevented the construction of some waste incinerators and the establishment of some landfills. In 1992, Germany had to export around 800,000 tonnes of municipal waste to France, due largely to the opposition of committed German environmentalists to the construction of waste management facilities, particularly waste incinerators" (Schmitt-Tegge 1994:441).

Germany is the leading example of a country where an ecological party has managed to gain significant electoral representation at the national and state levels. The Green Party was formed in 1980 out of a variety of smaller left-wing, peace, and ecological activist groups, mainly in response to the issue of nuclear energy and weapons. However, as with other Green parties, the German Greens are more than a single issue group. They offer distinctive positions on a wide variety of issues and encompass a broad range of different opinions. These differences nearly proved to be its undoing, and probably play a large, though not the only, role in explaining the decline of their electoral fortunes in the early 1990s. The Greens' first electoral breakthrough came in Land elections in the early 1980s. In 1983 they broke into national politics, polling 5.6 percent of the national vote, thus forming a parliamentary delegation of 27 deputies (Frankland 1995:29).[48] At the Land and local levels, the Greens achieved a high level of success. As Frankland describes it, "their Land electoral successes had noteworthy consequences: displacing the FDP as the third party, reducing the political power of the SPD, creating the possibility (in two cases the actuality) of SPD-Green coalitions and/or forcing new elections. Furthermore, some 7,000 Green local councillors had been elected, and in a number of major cities local Greens had become an important variable in power equations" (ibid.:30).

However in 1990, the first elections following unification, the West German Greens polled a mere 4.6 percent of the vote, thus losing all national seats. The Greens have rallied since then, regaining seats lost at the Land level, and reentering the Bundestag in 1994 with 7.3 percent of the popular vote in the October federal elections. The success of the Green Party has important implications for this analysis. First, it would not have been possible without institutional structures—most importantly, Germany's electoral system—favorable to the representation of environmental interests (or, indeed, many other interests) at the national and state levels. In many ways, the Greens were not successful in their aims of challenging and reforming "politics as usual" in the Federal Republic (Hager 1995). However, their influence in the policy sphere has been significant. Hager argues that "in a sense, the results have been impressive. Germany now has some of the most stringent environmental laws in Europe. The level of information of the average citizen is also quite high. Although grassroots activists failed in many of their protests against specific projects, their influence is evident in the central place environmental policy now occupies on the national political agenda" (ibid.:16).

Hence the Greens, and their constituency, became a force that could not be ignored in German politics. The mainstream parties responded to the Green challenge by placing environmental concerns high on their election manifestos, thus undercutting the Green Party's constituency (Héritier et al. 1994:56).[49] In October 1998, in an unprecedented victory, they achieved membership as full partners in the relatively unstable coalition government headed by SPD leader Gerhard Schroeder, a move that ultimately required the party to compromise on a number of its ideals. However, it made notable strides in reducing Germany's commitment to nuclear power.

The picture of state-society relations in the German case is very different from that of the British case. The high placement of environmental issues on national and state policy agendas not only discourages waste imports, but also ensures that Germany has one of the most comprehensive waste management policies in the OECD. A strong and vibrant environmental community has been able to find ways of ensuring that the government responds to their demands. This achievement is related primarily to the different points of access afforded to societal interests within the policy process. The electoral system enabled the rise of a Green Party, and the federal and judicial systems give a voice to other individual and group interests within German society. At the same time, the picture is far from rosy: the actual process of environmental policy formation still occurs within a relatively narrow policy community, and there still exists much opposition to the strongly technocratic modes of policy formation and implementation. In a sense, this has provided a focal point for environmental groups to organize around. It also leaves a large area of the policy process unexplored in terms of this analysis. Hence, in the next two sections I explore some of these themes.

The Politics of Risk: Technocracy and Government-Industry Relations
As in the U.K., the relationship between government and industry in the

process of policy formation in the Federal Republic is very close, and industry representatives play an important part in the formation of environmental policy and its guiding principles.[50] Some analysts also point to similar patterns of government-industry interactions in German politics more generally.[51]

Unlike the U.K., however, there is in Germany no pattern of strong and successful lobbying on the part of waste disposal firms to allow waste importation from abroad on a large scale, although there is some evidence that preferences for waste importation do exist on the part of both publicly and privately owned firms. In principle German policy is that wastes are supposed to be disposed of as close as possible to their point of generation (OECD 1993c:57), although in practice the Länder do in fact trade wastes among each other.[52] According to Pflügner and Götze, the highest importers by far are (still) the Neues Bundesländer —the former DDR—which accept imports from at least six other Länder (Pflügner and Götze 1994:9); Bremen, Lower Saxony, Schleswig-Holstein, and Hessen also accept wastes from other Länder, but rarely from abroad.[53] Hence there is some evidence that waste disposal firms, even those that are publicly owned, would like to engage in the exportation and importation of hazardous wastes (Rethmann 1990). However, institutional constraints and high disposal costs have made it extremely difficult for them to act on these wishes.

Several reasons why the waste disposal industry is less powerful in Germany than in Britain lend powerful support to an institutionalist explanation of German waste trading practices. One—the structure and management of the waste disposal industry—will be discussed in more depth in the following section. Others—discussed below—relate directly to differences in the processes of risk assessment and the degree of formalism in government-industry relations.

The German mode of policy implementation, with its roots in formalized legal principles, such as the precautionary principle, and its implementation of strict, across-the-board standards, has very much constrained German industry's freedom of action in comparison with British industry. Specifically with respect to the waste trade, strict standards of pollution control, high levels of monitoring, and an emphasis on using the best available disposal technology have resulted in some of the highest waste disposal costs in the world. This factor constitutes a significant trade barrier because it has made Germany an extremely unattractive destination for firms in other countries looking for potential recipients for their wastes. It has also contributed to the thriving illegal export business in Germany and to a number of waste exportation and dumping scandals in recent years.

In terms of government-industry relations, a much higher degree of formalism and state control is evident in Germany than in Britain. Some relate this to Germany's long-standing corporatist tradition, and others to the tradition of *Staatlichkeit*, a concept that encompasses both the importance of hierarchy in the German tradition—whereby the state lays down policy options to which interests respond (Héritier et al. 1994; Aguilar 1993)—and the formal incorporation of interests into the policy process.[54] Thus actors are constrained within a set of clear and publicly acknowledged guidelines and parameters that are key in understanding why German industry is not able to translate its preferences as effectively into policy outcomes as British industry generally is. In fact, Germany and Britain have very different traditions of state intervention. For example, in Germany the state has taken on the role of national protector in several respects. High levels of state intervention—top-down, or *von oben nach unten*—are accorded a high degree of legitimacy by different societal interests, in part via the grounding of policy in legally driven and constitutionally grounded principles (Weale 1992a:74). In Britain, on the other hand, a more pluralist notion of government controlled by public opinion has traditionally prevailed (although in practice, it has worked out otherwise).[55]

In recent years, there have been a number of indications that the policy community in Germany is opening up to include broader representation of societal groups (Blühdorn, Krause, and Scharf 1995; Aguilar 1993). One is the previously noted generational change among German bureaucrats. Another is the success of the Greens and the subsequent "greening" of established party platforms, which has undercut the role of industry in the policy process and opened up the deliberation process to a variety of different environmental organizations.[56] Many leading German environmental groups are changing their tactics, exchanging street-level protests for high-level lobbying—a professionalization of activities (Blühdorn 1995).

Germany's technocratic and statist mode of environmental regulation has therefore affected processes of risk assessment in two important ways. First, a reliance on scientific modeling and stringent standards has led to more conservative risk assessments on the part of policymakers. Second, industry's freedom of action has been considerably restrained by Germany's statist tradition of regulation. Furthermore, it seems that industry is willing to accept both the constraints imposed on it and the role it plays in the policy process. Adding in the increased importance attached to public opinion on environmental matters, it becomes possible to identify two main routes through which waste imports are inhibited. First, due to stricter regulatory standards, the costs of waste disposal are much higher in Germany than in other OECD nations. Second, given both public opposition and the risks posed by waste importation, allowing large-scale importation of wastes runs entirely counter to Germany's conservative approach to environmental risk. In fact, waste importation barely figures as a part of the debate in the German environmental policy arena.

Relative Absence of Principal-Agent Dilemmas

A key reason that levels of waste importation in Great Britain are so high is that the diffusion of waste regulation and disposal responsibilities has created a positive incentive for local authorities and firms to engage in waste importation, while central government actors find it extremely hard to control these practices. In Britain a three-level "principal-agent" chain exists, consisting of central government actors, local authorities, and the hazardous waste disposal industry. Incentives for engaging in waste importation are provided not only by the lack of policy coordination, but also by the financial benefits available to local governments and industry actors that engage in waste importation as a way of raising revenues/profits. In contrast, the scenario in the Federal Republic considerably reduces the incentives for the different actors to engage in or permit waste importation from abroad.

The waste management industry in Germany is structured very differently from its British counterpart. Practices vary by Land: some rely on privately owned, decentralized facilities, while in others the waste management sector is highly integrated and publicly owned and managed. In most cases, however, there is a high level of federal and

Land government involvement, for example, through subsidies and other forms of support. This means, as I argued in chapter 3, that the waste management sector on the whole is more accountable to government and societal actors for its behavior, and hence less free to engage in what might be interpreted as antisocial goals, such as waste importation. Furthermore, they are cushioned from the sort of economic uncertainty that characterizes the operating environment in which British firms have found themselves. In addition, the restrictions placed on movements of waste by the federal and Land governments, as well as barriers to entry into the German market, has meant that German waste disposal firms have fewer international or multinational connections than their British counterparts.

Finally the relationship between federal and Land governments in Germany is not (nor can it be) characterized by the same dynamics that prevailed in Britain throughout the 1980s. Each level of government operates and interacts within a constitutionally protected framework of rights, responsibilities, and financial arrangements. In terms of the waste trade, Land governments have more powers of control than British local authorities and little incentive to allow waste importation by firms in their jurisdictions. Indeed, they only do so when it appears that there are serious capacity shortfalls.[57]

Concluding Analysis

Germany's system of environmental regulation exhibits certain central characteristics that have determined its risk-averse behavior toward the international waste trade. In fundamental ways, its structure and style of environmental regulation are diametrically opposed to those in Great Britain. Not only is Germany a net exporter rather than importer of hazardous wastes, but also the basic contours of the policy debate surrounding the generation and disposal of hazardous wastes are sharply divergent in the two countries.

First, the relationship between government, industry, and societal actors has been configured in very different ways by institutional structures. German intergovernmental relations (between the federal and the Land governments) are much more structured than British intergovernmental relations: carefully delineated constitutional roles and guaranteed sources of revenue have placed the Länder in a more secure position than the British local authorities. At the same time, the Länder are bound by law to discharge certain centrally determined obligations toward the environmental protection of their citizens. For their part, individual citizens and environmental interest groups have many more points of access to the policy process—through the electoral system and the courts, and at different levels of government—than their British counterparts, and they have used this access to good effect.

Second, Germany's highly technocratic and, indeed, rigid mode of policy implementation relies on the precautionary principle (*Stand der Technik*) and the polluter-pays principle as guides to industrial practice. This has fostered an extremely risk-averse attitude toward allowing in pollutants from abroad and raised waste disposal costs over the equivalent median prices in other countries, especially in contrast with Britain's more flexible, firm-oriented implementation practices.

Furthermore, there is evidence that the policymaking process is opening up to include the increasingly professionalized elements of the German environmental movement in day-to-day government decision making. This, as Héritier et al. (1994) have put it, means that actors in the German system, unlike in the British system, are not in a "zero-sum" situation: the emphasis on consultation and consensus according to formal and well-defined rules means that no one voice is able to dominate policy outcomes. Industry, too, is under more constraints than exist under the British system. A long tradition of state intervention in the German hazardous waste disposal sector and the high legitimacy accorded to state intervention in curbing industry's power are in sharp distinction from Britain, where the large waste disposal firms have been able to gain the upper hand in a policy process both secretive and informal.

When the waste trade has come up on the political agenda in Germany it has been in terms of legal or illegal waste exportation, not importation. Within Western Europe, the official German policy is one of self-sufficiency at the national level (Pflügner and Götze 1994). Germany has, however, entered into a number of cooperative agreements—for example with France, Belgium, and the Netherlands—to allow waste trading (in both directions); such agreements appear to be, in practice at least, an endorsement of the proximity principle of waste disposal.[58] Its exports of waste to further-flung OECD nations also appear to be an official endorsement of such practices. In addition, Germany is a signatory of the relevant international agreements that ban the export of wastes from developed to less-developed countries. However, as examples of illegal exportation show, the German government has not fully enforced such bans. In EU forums, Germany continues to oppose a ban on the export of wastes for recycling purposes.

German national policy is aimed at reducing waste production at the source and increasing levels of recycling and reuse of wastes in the product cycle. That the Germans have undertaken a vigorous and, according to early data, successful campaign to reduce the generation of hazardous wastes—regulation at or before the start of the waste cycle—is entirely consistent with their underlying philosophy of pollution control.

Recent moves toward a closed-circle economy (the *Kreislaufwirtschaft*) are a logical extension of German environmental policy and practices. This legislation was sparked primarily by three external factors that forced the German government to confront the issue of high levels of hazardous waste exportation.[59] First, German unification effectively removed what had been a fairly easy solution to the problem—dumping wastes over the border into the DDR—and meant that the Federal Republic now had to bear the costs of cleaning up contaminated sites in the new Länder. Second, the outcry in France in 1992 following the dumping of dangerous medical wastes led to a subsequent (and temporary) ban imposed by France on most German waste imports. Finally, the EU and the international community increasingly forbid or restrict waste exportation by their members.

Proposed in 1992 and implemented in late-1996, the Kreislaufwirtschaft legislation has important implications for German hazardous waste management.[60] First, waste definitions now comply with EU regulations and include wastes destined for recycling (previously categorized as economic goods).[61] Second, waste generating firms are required to reduce waste production and increase recycling, either via energy production or reuse of wastes as secondary raw materials. Finally, the legislation imposes strict producer responsibility for all wastes generated and tightens up licensing and monitoring processes for waste exportation.[62] Early results of this initiative—for example, anticipatory action by firms—have been impressive (Czech 1996).[63] It is expected that the generation of hazardous wastes for final disposal will continue to decline, which should have a corresponding effect on waste exportation—especially if disposal prices fall as a result of spare capacity opening up in the disposal industry. However, it is by no means clear that Germany will begin to import wastes from abroad, as argued by Rethmann (1990).[64]

In the face of internal and external pressures for change in policy practices, such as European integration and unification with the former East Germany, the German system of environmental regulation has remained remarkably constant.[65] It tends to adapt to changing economic, social, and political conditions rather than reform in reaction to them, and there is little evidence that these have had much effect on the German style and structure of environmental regulation.[66]

The nature of government-industry relations has played a large role in this relative constancy. There appears to be a remarkable acceptance of government regulations on the part of industry actors, and the government (Bund and Land) has in recent years begun to show more flexibility in its application of pollution control mechanisms to individual firms. Examples of this flexibility include more extensive reliance on voluntary agreements between firms and between government and firms, and a growing emphasis on producer responsibility for causing environmental harm, as evidenced in the controversial Packaging Ordinance of 1991, which obliged companies (both national firms and importers) to collect any packaging used in the shipping of their products to the point of sale.[67] In fact, many argue that the economic growth/environment tradeoff is viewed very differently in Germany than in other industrialized countries.[68] For example, Germany leads the world in the production of environmental goods and services, and up to a quarter of total investment is toward environmental protection (Randlesome 1994; OECD 1996).

Thus in Germany, as in Britain, hazardous waste management practices are at a crossroads. Public opposition to the construction of new facilities and heightened international awareness of, and opposition to, Germany's waste exportation practices have led to an even more rigorous approach to waste regulation. The implementation of the Kreislaufwirtschaft (Closed-Circle Economy) Ordinance has shifted the regulatory focus away from the waste disposal sector onto the actual generators of wastes. The German regulatory system's ability to adapt to changing circumstances (as opposed to attempting any fundamental reform) demonstrates that institutional features have even greater explanatory power in this case.

What the future holds is another matter. Some trends—for example, the decline in capacity utilization and the growing role of the private sector in waste management—point toward a possible growth in demand on the part of the waste industry to allow importation. These trends could be strengthened should the EU waste regime move toward allowing wastes to be exported to countries with the best facilities. However, as argued above, this looks unlikely, especially given the powerful influence societal actors wield over the policy process, and the application of the precautionary principle as a basic tenet of German environmental regulatory philosophy. In the final chapter, I discuss the implications of the basic and persistent differences between Germany and Britain and other OECD countries for the international regulation of the waste trade.

6
The Waste Trade and Environmental Regulation in France, Australia, and Japan

In this chapter I examine whether the institutional model developed in earlier chapters has general applicability to cases other than Britain and Germany by extending the empirical analysis to three shorter cases: France, Australia, and Japan. I examine the extent to which the hypotheses concerning regulatory style and structure fit each country's propensity to import hazardous wastes and, to a limited extent, test the regulation hypotheses against each other. These three cases are an interesting test of the adaptability of this model across different industrialized countries. On the whole, they support the central claim of this book: that differences between national systems of environmental regulation explain why some countries are legal net importers of wastes and others are not. Ideally, other cases might be examined as well; however, these three are presently the only other OECD countries for which reliable data and/or analysis is available over a number years regarding their legal hazardous waste importation habits and their styles and structures of environmental regulation.[1]

Results for this chapter are summarized in table 6.2. France, the world's largest legal importer of hazardous wastes, is a close match for Britain: it has a decentralized waste management and regulation structure and a closed policy style. Australia and Japan are useful cases in that they are not members of the EU; hence their inclusion allows me to examine waste trading practices in countries with different transnational links. Australia (somewhat surprisingly), of all the cases examined in this work, has the most centralized waste management structure—much more so than Germany. Finally, Japan is an anomaly. As table 6.2 shows, it matches the British model right down the line; however, its

Table 6.1
Net imports (exports) of hazardous wastes by Australia, France, and Japan, 1989–1993

	Imports (tonnes)	Exports (tonnes)	Net imports (exports), (tonnes)
1989			
Australia	0	500	(500)
France	n.d.	n.d.	n.d.
Japan	5,125	40	5,085
1990			
Australia	0	1,000	(1,000)
France	458,128	10,552	447,576
Japan	397	0	397
1991			
Australia	0	3,200	(3,200)
France	636,647	21,126	615,521
Japan	n.d.	n.d.	n.d.
1992			
Australia	0	275	(275)
France	512,150	32,309	479,841
Japan	n.d.	n.d.	n.d.
1993			
Australia	0	0	0
France	324,538	78,935	245,603
Japan	n.d.	n.d.	n.d.

Source: OECD 1994a, 1997a, table 1. n.d. = no data provided. Australian data refers to permits for final disposal.

attitude toward waste importation displays a high degree of risk aversity. In the third section of this chapter I discuss why this is, focusing on the pivotal role played by a series of environmental crises in the 1950s and 60s in Japan in shaping both its regulatory system and the attitudes of government and business actors.

As well as supporting the main argument, these three cases demonstrate the variety that exists in modes and practices of environmental regulation across OECD countries. In each case it was possible to identify prevailing styles and structures of environmental regulation; however, certain national differences show just how many ways these fit

Table 6.2
Main variables and findings

	France	Australia	Japan
Structure of waste management and regulation	Diffuse	Centralized	Decentralized
Access to policy process	Closed	Open	Closed
Mode of policy implementation	Rigid	Rigid	Flexible
Waste importation propensity	High	Low	Low

Notes:
Structure of waste management and regulation = allocation of regulatory authority between national and subnational authorities, number of regulatory authorities, and structure of waste disposal industry (public or private; degree of competition).
Access to policy process = composition of environmental policy community plus a country's political opportunity structure (e.g., electoral system, judicial review).
Mode of policy implementation = use of standards; monitoring and enforcement mechanisms and principles.
Waste importation propensity = a country's status as net importer of wastes, measured over time.

together to affect risk propensities. For example, Australia has a very different sort of federalism than Germany, and France's administrative tradition differs from Britain's in ways that have made providing effective environmental protection in France extremely problematic.

France

In many ways France is a "big unknown" in studies of comparative environmental policy. Its EU counterparts such as Britain, Germany, the Netherlands, and Sweden have received much more attention in the literature. There are several reasons for this absence. French environmental policy practices do not provide as sharp a contrast as do Britain and Germany, nor have the French developed a comprehensive or innovative policy program, as has the Netherlands with its National Environmental Policy Plan. A related factor is the widespread perception that environmental issues are not as highly placed on the French political agenda as they are elsewhere, and that environmental consciousness on the part of the general public is comparatively low.

France is the world's largest legal net importer of hazardous wastes. According to the OECD, in 1990 France's net imports reached 447,576 tonnes; in 1991 they were reported at 615,671 tonnes, and in 1993 at 482,000 tonnes (OECD 1994a, 1997b; see also table 6.1).[2] This risk-acceptant behavior with respect to waste imports is reflected in other policy areas, most notably the use of nuclear energy, on which France relies to supply much of its domestic energy needs. On grounds of France's size and population, its hazardous waste importation practices do not, on an intuitive level, provide the same sort of puzzle as Great Britain's. However, the extent to which its system of environmental regulation and its preference configuration of main stakeholders in the trade parallel Britain's makes France a good case for this analysis.[3] The French model mirrors the British in many respects. It is a unitary state with a strong tradition of statist intervention, and it provides low-cost, privately owned waste disposal facilities—regulated primarily at the local level—for potential exporters that are geographically proximate to Germany and Switzerland, its main waste trade partners. It also has, as shall be seen, a very closed policy style.

Center-Local Politics and Regulatory Structure in Fifth Republic France
Like Britain, France's structure of environmental regulation, and hazardous waste management in particular, is decentralized to the point of diffusion. Regulatory responsibilities are allocated among different ministries and levels of government, whose roles are considered by many observers to be both uncertain and highly ambiguous. In addition, France's hazardous waste management industry is privately owned and characterized by a high degree of competition among the main firms.

This characterization admittedly goes against the most commonly held view of the French state: that it is highly centralized and that the country is run from Paris in a hierarchical and interventionist manner. This may hold true for some issue areas, but it is not the case in the environmental arena.[4] In fact, the contradiction between the structure of French political administration at the center and the nature of hazardous waste management and environmental regulation more generally has greatly hampered efforts to deal effectively with environmental degradation. This contradiction serves to exacerbate the coordination and monitoring problems already arising from its decentralized structure of hazardous waste management. As one analysis puts it, the key question facing French environmental policymakers is "how to establish an administrative or policy-making apparatus capable of dealing with an inherently transversal issue, the environment, within a rigidly segmented and hierarchical policy-making structure" (Bodiguel and Buller 1994:104). These structural factors interact along the lines predicted by the institutional model I have presented, thus explaining France's highly risk-acceptant attitude toward the hazardous waste trade.

Center-Local Government Relations The basic political institutions of the French Fifth Republic differ in many significant ways from the parliamentary and presidential systems of the rest of the OECD. The Fifth Republic came into being in 1958 at the behest of General de Gaulle and was intended to liberate France from the instability and weakness of continual government change, which was seen as the result of years of parliamentary dominance under the Third and Fourth Republics. De Gaulle established a system where the President dominated, although counterbalanced by a bicameral parliament.[5] This concentration of power is reinforced by the fact that France is a unitary state, although with a complex (and increasingly powerful), multitiered system of local government.

The French model of public administration, described by some as "attractive to tidy minds in untidy countries" (Wright 1994:116), is based on a Napoleonic model, variously "statist, powerfully centralized, hierarchically-structured, ubiquitous, uniform, depoliticized, instrumental, expert and tightly controlled" (ibid.)—an instrument of the state and relatively impervious to societal pressures. Practice, of course, deviates from theory—Wright points to the growing politicization of the bureaucracy and increased demands for performance evaluation as two of the main changes of recent years. However, it is still doubtful whether the Paris élite are able to handle complex, highly differentiated, and multilevel issues such as environmental policy.

France has a complex, five-tiered system of local government, and, as is in most OECD countries, subnational authorities play an important role in the implementation and, increasingly, the initiation of environmental policies, including hazardous waste management. The most important units of authority are the 22 regions, the 96 departments, and the multiple municipalities. Of these, the regions are the most ill-defined in terms of responsibilities, as it is only recently (since 1982) that they have been institutionalized in their current form (Douence 1994).

Center-local relations are shaped by a number of factors unique to the French political system. Local authorities—particularly departments—are seen as the "arm of the state" in local communities: "the functions of government are not provided by either local authorities or central government as in Britain, but in large part by deconcentrated offices of the central administration operating principally at the level of the department to form a hierarchy whose apex is in Paris" (Hunt and Chandler 1993:55). However, an informal network of "patron-client" relations ensures that influence runs both ways: "more than 80 per cent of national deputies hold an office in local government," thus placing them in a position of influence at the center (Hunt and Chandler 1993:67; Mény 1996:132). Local representation is increased at the center by the predominance of local interests in elections for the Senate.

In recognition of the need for greater local government autonomy and responsiveness to local needs and conditions, the French system is, if anything, moving more toward decentralization and possible federalization of intergovernmental relations. The 1982 regional reforms reorganized not only the structure of local government but also its responsibilities, according for the first time executive power to regional leaders (Douence 1994; Bodiguel and Buller 1994; Mény 1996). The implications of these changes for environmental policy, most of all their current ambiguous state, are discussed below.

Structure of Environmental Regulation The French system of environmental regulation evolved within the political context outlined above, in parallel with the growing decentralization of French politics. France has traditionally been viewed by its European counterparts as a laggard in the development of an effective environmental policy infrastructure. The first law regarding the disposal of industrial wastes—including hazardous wastes—was enacted in 1975 (followed by a second law in

1976), and France is now implementing a comprehensive national waste management plan that came into effect in 1996.

France's national Environment Ministry was established in 1971. Its powers are weak and in many cases subordinate to other ministries and governing bodies, including the President's office and authorities at the regional and Department level: "its exact functions have remained largely ill-defined and its form has been in a virtually perpetual state of reorganization" (Bodiguel and Buller 1994:98).[6] It lacks, for example, the sorts of links with powerful pressure groups that enable a ministry to improve its status (Stevens 1996:299), and its finances are severely curtailed in comparison with other ministries—in 1990 it received a mere 0.058% of the national budget (Héritier et al. 1994:160). Héritier et al. also note that many of the key government-industry negotiations over environmental issues occur under the auspices of the Ministry of Industry.

There are also several advisory bodies at the national level—in the case of waste management, the National Agency for the Recovery and Elimination of Waste (ANRED, established in 1975). Although ANRED's functions are primarily consultative and it lacks regulatory powers, it does have the power to allocate grants and low interest loans to waste minimization and recycling/recovery projects (Davis, Huisingh, and Piasecki 1987:28–29).

As in Britain and Germany, powers of policy formation in France are concentrated at the center, while implementation responsibilities, including the authority to issue waste importation permits (Biod, Probert, and Jones 1994:22), are delegated to lower levels of government (Brusco, Bertossi, and Cottica 1996:122). Local and regional governments, however, often do not have the resources or capacity to control waste importation (Biod, Probert, and Jones 1994), nor is there much evidence of harmonization of regulatory practices among different administrative units (OECD 1997b).

The French system of hazardous waste management has been significantly affected by the lack of clarity surrounding the role and powers of the regions following the local government reforms of 1982. As Bodiguel and Buller point out, "the creation of the regional tier has coincided with the emergence of French environmental policy" (Bodiguel and Buller 1994:92). In many ways, the regions are ideally situated to take on environmental administration, not only in terms of size but also as an extension of their earlier role in local economic planning and development. For example, under the 1992 Waste Management Plan, seen by many as formalizing existing practices, regions are given responsibility for the management of industrial waste disposal (Souet 1993).[7] The regions are, however, in something of a double bind with respect to effective environmental management. They are largely viewed as artificial constructs, removed from the day-to-day level of French political life embodied in the communes and departments. Furthermore, like British local authorities, their financial resources do not match their broadened responsibilities, which in turn lack the specificity necessary for effective administration.

Hazardous and Industrial Waste Management Industry in France
There are striking similarities between the British and French systems of hazardous waste management, in terms of basic characteristics of the industry and modes of disposal (Litvan 1995). According to a recent analysis, France produces roughly 150 million tonnes of industrial wastes per year, of which roughly 18 million tonnes are classified as *déchets spéciaux*, requiring special handling (Pecqueur 1992:24–5).[8] A large portion of domestically produced wastes are disposed of or stored onsite in often poorly regulated and unsafe facilities (Souet 1993). Of wastes disposed of in commercial facilities, including imported wastes, most are sent to landfill, and the remainder incinerated or chemically treated prior to landfill. In addition to 484 landfill sites (Stanners and Bourdeau 1995:351; *Haznews* 93, December 1995:18), France has 10 incineration facilities, 8 physiochemical centers, 8 evapoincineration plants, and 7 treatment centers for soluble oils; many of these are situated around France's borders (Pecqueur 1992:27–28). Owing to its size and low population density, it is commonly recognized that France does not have the capacity problems faced by the rest of Europe, and its rate of waste incineration is one of the highest. In particular, it is one of the few countries with spare capacity for the treatment of PCBs (OECD 1997b:98), although it also has a huge problem with abandoned and contaminated landfill sites.

As in Britain, most of France's treatment facilities are run by private companies, although these are heavily subsidized by government agencies, which in turn has helped waste disposal costs remain well below the European average. The French waste disposal industry is highly concentrated, with two firms—Compagnie Générale des Eaux (Onyx) and Lyonnaise des Eaux (SITA)—controlling about 20 percent of the market each, followed by two medium-sized firms and a collection of small enterprises (Brusco, Bertossi, and Cottica 1996:116). Both the main conglomerates have extensive interests abroad and are highly diversified in terms of their activities.[9]

Government involvement in this industry has in the past consisted primarily of financial measures—subsidies to cover the costs of improvements (Leroy 1987:146–7) and, more recently, a landfill tax similar to that imposed by Britain. Although initially set at levels too low to affect waste disposal routes, the landfill tax is in the process of being raised to a more significant level (see below). The combination of government subsidies and low-cost disposal facilities has contributed to France having among the lowest disposal costs of the more industrialized European countries; these costs, however, have risen sharply in recent years. *Haznews* relates this to the "monopolistic trend" in the French waste disposal industry, although a significant drop in waste importation has not accompanied this rise in costs.[10]

French Style of Environmental Regulation: Access and Implementation

The French system tends to exclude societal interests, particularly from access to the policy institutions at the center. In terms of actual policy-making procedures, the traditionally statist style—which emphasizes hierarchy, the powerful position of the President, and little local government autonomy—precludes open discussion and debate over policy initiatives (van Waarden 1995:361). For example, regulatory powers are largely removed from the legislature, and discussions tend to occur within committees of public officials and scientists "with little open controversy and with no incentive to explain their actions or to change their regulatory priorities" (Brickman, Jasanoff, and Ilgen, 1985:305). However, Brickman, Jasanoff, and Ilgen also note that there is a high degree of public acceptance of this process of regulation (ibid.:308).

Unlike Britain, the policy process in France is closed to industrial interests, at least at a formal level, on the basis that such activity goes against the role of the state as a neutral director of social and economic development: "[The French authorities] are much less enthusiastic about involving private interests in public policy as they fear that the particularism of these interests will threaten the 'national interest' for whom they themselves stand guard" (van Waarden 1995:344). Although, there is a consultation role for peak associations in some areas of policy, "the French government gives industry a less privileged position [than in Britain or Germany] in the [environmental] policy process. French civil servants rely more on a variety of technical and advisory bodies than industry associations" (Heidenheimer, Heclo, and Adams 1990:325).

At the same time, there is evidence both of informal contacts between government and industry representatives and of influence exerted by industry at lower levels of government. Both of France's big waste disposal firms are documented as having close relations with the EU and the national government (Brusco, Bertossi, and Cottica 1996). The French government has shown a willingness to support French waste disposal companies in their dealings with other countries on a number of occasions. Local politicians, too, are an easier target for polluting firms that wish to continue their activities in the face of regulation. Heidenheimer, Heclo, and Adams cite the case of Marseilles in the 1970s, where the Mayor not only had entered an alliance with local industrial firms but also held powerful positions in the national government: as a result, the firms remained largely untouched by regulations (Heidenheimer, Heclo, and Adams, 1990:338).

Broader political structures, including the electoral system, further contribute to a closed opportunity structure on the input side.[11] On the output side, several analysts argue that France exhibits a degree of openness, especially as the recent government decentralization has brought implementation decisions "closer to the people" (e.g., Héritier et al. 1994:147–148); however, this process is prone to the ambiguities of responsibilities and legitimacy noted above. In addition, the role of the judiciary as a means for citizens to force action on the environment is severely limited.[12]

Under the Fifth Republic constitution, the electoral system was designed to increase political stability and to foster a bipolar Left-Right party system. In the early years of the Republic, all elections were held on a two-round basis on the principle of "choose on the first, eliminate on the second." More recently, this system has undergone a series of reforms, and currently only the Presidential election is held under the two-round format. At municipal, departmental, regional, and European levels, a modified system of proportional representation is used, and at the parliamentary level proportional representation was used from 1982 until 1986, when the two-round system was reinstated (Mény 1996:105–6). The main result of this series of reforms has been a centrifugal effect on the party system at all but the presidential level, allowing "alternative" parties, such as Les Verts and the National Front, to gain representation at local levels of government and in the European Parliament.

In terms of France's mode of policy implementation, again, the statist tradition comes into play despite informal ties between government and industry (van Waarden 1995:361; Héritier et al. 1994:170). Environmental officials have a wide range of powers and policy options they can apply to firms, which in turn are subjected to environmental impact assessment procedures. Some argue that these procedures, even from the German perspective, seem somewhat draconian (Héritier et al. 1994: 169–170). Litvan (1995) concurs, arguing that the French could learn from the more flexible British approach to waste management.

Stakeholder Relations: Institutional Constraints and Risk Acceptance

The question now becomes how the above institutional constraints—regulatory structure, access to the policy process, and mode of policy implementation—have shaped relations between government, industry, and societal actors to enable waste importation on such a vast scale by private concerns. Two factors are key. First, France's administrative tradition is one of high centralization coexisting with an allocation of responsibilities to ill-equipped local governments. This arrangement is inappropriate, and possibly dysfunctional, for managing an issue as complex, dynamic, and multilevel as the importation of hazardous wastes and their adequate disposal. Recent changes such as the introduction of the regional tier of government have left regulatory responsibilities ill-defined and confusing, and so have done little to improve the situation. The French system is possibly even less coordinated than the British model. For example, the French government has been extremely slow in implementing a national waste management plan (Litvan 1995; Souet 1993). The French waste disposal industry has taken advantage of this situation, and has been able to import wastes from neighboring countries for disposal at very cheap—usually landfill—facilities. Second, while several factors demonstrate a reasonably high level of environmental awareness and concern among French citizens, there are very few channels for registering public opposition to waste importation via electoral politics, pressure groups, or the courts.

Although the French environmental movement is nowhere near as diverse, populous, or influential as Germany's—there are few environmental pressure groups operating at the national level—France has a strong record of vocal and active localized groups organized around environmental issues. Les Verts, France's main Green party, maintains a consistent level of support both in opinion polls (14% between 1989 and 1992) and in local and European elections. This consistency implies support beyond a mere protest vote.[13] Probably the biggest environmental group is the Fédération Nationale des Sociétiés pour la Protection de la Nature, which is strongest in Rhone-Alpes, where there was an early nuclear controversy. Many of the regions where environmental protest is strongest tend to be on France's peripheries—the regions where most waste disposal facilities for dealing with imported wastes are located.[14]

Most relevant to the waste importation issue, high levels of local opposition to the dumping of toxic German hospital wastes in 1992 prompted a government ban on all waste imports from Germany.[15] According to figures issued by the German Environment Agency in 1996, the ban proved temporary; however, the legislative result of this controversy was a reinforcement of the powers of local authorities to stop wastes from entering the country. Under decree 92–798 of August 18, 1992, "imports of wastes are prohibited except if they originate from an EU member state and conform to a waste elimination plan or if such imports result from an agreement between France and a non-EU member state" (Biod, Probert, and Jones 1994:22). These caveats thus open the way for a large number of exceptions and, interestingly, also mean that France has effectively sidestepped EU and UNEP regulations on the importation of wastes from other EU member states and from outside Western Europe.

France: Concluding Discussion

As the above discussion has shown, the institutionalist argument can be successfully applied to the French case. Like the United Kingdom, France has a highly decentralized environmental regulation structure—decentralized to the point of diffusion—within the context of a unitary political system. The Environment Ministry is weak, and implementation powers are scattered among a collection of regional governments and agencies. In turn, this has hampered government efforts at policy reform and has granted the main waste disposal firms a good deal of leeway in their actions. Furthermore, France's closed policy style has precluded the involvement of interests that oppose waste importation, except in a few egregious cases (such as the German medical waste scandal). The one main point of difference between the French and British systems in terms of this model is, however, that while Britain has a flexible mode of implementation, France's is best described as rigid. At the same time, the configuration of preferences of industry and societal actors resembles that existing in the United Kingdom and Germany, with industry favoring imports and public opinion and environmental groups opposing them.[16]

Thus the combination of factors specific to the French case and the high degree of correlation with the British case support the hypotheses put forward in chapter 3 regarding the regulatory determinants of waste importation. As in Britain, recent changes in French government strategy and regulatory practices might have an impact on its waste importation propensities. A ban on sending untreated wastes to landfill is due to come into effect in 2002 (Litvan 1995).[17] This in turn has stepped up the search for viable alternatives to landfill, including the construction of incineration facilities capable of energy generation. Plans have also been published to implement a more comprehensive waste management strategy that would formalize the allocation of planning functions at the national level and implementation at the regional level (OECD 1997b:94). Buller (1998) questions the impact Europeanization is likely to have on France's environmental policy: along with Britain, it has been an outlier in this area of integration. Probably the most important question regarding environmental policy development in France is whether the country will move toward the more well-defined and coordinated German federal system or the more diffuse model that has characterized British environmental policy for so long.

Australia

Many factors point toward Australia being a waste importer. It is thought by many to have plenty of practically deserted or sparsely populated land, and its neighbors include many of the most densely populated and rapidly industrializing countries of Asia, most of which do not have adequate waste disposal facilities or infrastructures. It has also been known for having a strong pro-development lobby, which would encourage such entrepreneurship. However, Australia is not a waste importer, and in fact it has gained some notoriety as a waste exporter (see table 6.1).[18] The institutional model effectively captures the reasons for this. In particular, analysis of Australia's regulatory structure and style explains not only why it does not import wastes but also why the government has so far been unable to guarantee an adequate treatment infrastructure. Furthermore, like Germany, Australia is a federal nonimporter of hazardous wastes—hence I also explore in this section whether pertinent conditions in the Federal Republic are similar to those in Australia.

Australia exports hazardous wastes—especially of the "intractable" variety—because it lacks the sorts of facilities (most notably, high temperature incinerators) to dispose of them in a proper manner (McDonell 1991).[19] Such wastes, therefore, have been shipped to the United Kingdom and to France.[20] Remaining intractable wastes have been stored in facilities in and around Sydney, the capital of Australia's most heavily industrialized state, New South Wales (NSW).[21] Other hazardous, but nonintractable, wastes have been sent to landfill or dumped in the oceans (Lipman 1990:288).

During the late-1980s and early-1990s, a commission appointed by the commonwealth (federal) government located and attempted to begin construction of what was intended to be a national waste incinerator on a site close to the border between Australia's two most industrialized states, New South Wales and Victoria. However, extremely high levels of public protest followed, and as a result the project was halted (McDonell 1991). Waste importation and stockpiling have continued since then, as have attempts to find alternative solutions to the problem as waste exportation becomes less viable as a disposal option. These alternatives include exploring new sites or networks of smaller sites, developing waste management and recycling plans, and examining alternative strategies of waste management (Beder 1991). This in turn has led to more consideration of alternative modes of waste disposal—bioremediation, for example—than in the other countries discussed in this book.

In this section I examine the institutional basis of Australia's relatively risk-averse attitude toward waste management and the waste trade, focusing in particular on how evolving state-center relations in the area of environmental policy and channels of access to the policy process in Australia have shaped this outcome. I suggest that these factors are responsible for Australia's lack of waste importation, as well as its exportation practices, for two main reasons. First, hazardous waste management and regulation are relatively centralized: the federal government has control over the issuing of import and export permits, and disposal facilities are concentrated in two states. This greatly simplifies the coordination and principal-agent problems facing regulators in other countries (e.g., Great Britain and France). Second, NIMBY sentiments have been able to find expression in the policy process in ways absent in many waste importing countries—hence the absence of a centralized incineration facility in Australia.

Regulatory Structure: Commonwealth-State Relations and Regulatory Agencies

Australia's federal constitution, enacted in 1900, makes no mention of environmental issues.[22] Australia is made up of six states and two commonwealth territories administered by the commonwealth (or federal) government.[23] Until the early-1970s, environmental issues were seen primarily as falling under the authority of the states, most of which set up their own environmental protection agencies modeled on the U.S. EPA (Christoff 1994:350).[24] However, with the election of the Labor government under Prime Minister Gough Whitlam, the commonwealth government began to play an increasingly active role in issues of pollution, conservation, and resource management.[25] This ushered in a period of conflict and recalibration of the division of state-center responsibilities vis-à-vis the environment. According to one analysis, this period ended with the recognition that Australian federalism is not coordinate, where federal and state governments have separate spheres of authority, but concurrent, whereby states and center are recognized as having overlapping responsibilities in this issue area (Kellow 1996:135). However, institutions reflecting this realization only began to be developed in the early-1990s (ibid.:137).

The original pattern of environmental powers resting with the states has led to an allocation of environmental policymaking responsibilities very different from that in the other cases examined so far. In those cases policy formation occurs at the central level of government, and implementation powers are delegated to constituent states or local authorities. In contrast, Australian states historically have had complete jurisdiction over environmental policy issues within their boundaries—until recently, institutionalized mechanisms for dealing with environmental problems crossing state boundaries were weak or nonexistent.[26] However, the combination of activist Labor Party governments in the 1970s and 1980s, several high-profile environmental conflicts, and the beginnings of international environmental governance triggered a change in this pattern. The Franklin River Dam conflict provided a key focal point for a newly invigorated federal environmental initiative (Hutton and Connors 1999:158–164). Protests over plans to turn an unspoiled piece of the Tasmanian wilderness into a hydroelectric dam had mobilized thousands of activists. 1983 was also the year of a General Election, and the Labor Party, under the leadership of Bob Hawke, was able to make the protection of the Franklin River a central plank in their campaign. Their victory, along with the support of the Australian High Court, made it possible to declare the endangered area a World Heritage site and paved the way for greater federal intervention in states' management of the environment through the creation of several national environmental agencies.[27]

The main agencies at the center dealing with environmental issues are the Department of the Environment, Sports, and Territories (as it has been known since the 1993 elections), the Department of Primary Industries and Energy, and the Aboriginal and Torres Strait Islander Commission. The more recently established commonwealth Environmental Protection Agency plays a planning and consultative role (Kellow 1996:136).[28] The environment ministry ranks fairly low in the ministerial hierarchy, although, as I discuss below, it has managed to exert a fair amount of influence in recent years.

Following the Franklin River Dam court case, state-center relations continued in this pattern of confrontation over issues such as the preservation of tropical rainforests, uranium mining, and logging in different parts of the country (Papadakis and Moore 1994:341). All parties involved found this ad hoc, crisis-driven approach unsatisfactory, resulting in efforts to set up institutions to mediate these conflicts, of which the commonwealth EPA was one. Under the leadership of Prime Minister Bob Hawke (1983–1991), certain cooperative arrangements—such as the 1992 Intergovernmental Agreement on the Environment (IGAE)—and more specific arrangements—regarding the Great Barrier Reef, for example—began to emerge (Saunders 1996:74). However, Saunders cites the Biological Control Act of 1984 as the only example of the commonwealth enacting "model or 'template' legislation," the pattern of divided policy formation and implementation used in Germany (ibid.:74).[29]

Despite the relatively decentralized structure of environmental regulation overall, hazardous waste management in Australia is in fact relatively centralized. Two reasons explain this. First, most waste generating industries and waste disposal plants are located in only two states, New South Wales and Victoria. Sydney and Melbourne, their respective capitals, alone account for 70 percent of Australia's industrial production. Therefore "hazardous waste management in Australia is managed by these two state governments and the commonwealth (federal) government in Canberra" (McDonell 1991:33). Australia's waste disposal industry is for the most part privately owned and run (Robinson 1990), and there is some evidence of involvement in the Australian market by the large multinationals, such as Waste Management International. It is, however, small and relatively lacking in political clout.

The second reason relates to the constitutional division of powers and pertains especially to the international waste trade. The commonwealth government was able to assert its authority at an early stage over international environmental issues, particularly via its power to regulate trade and commerce with other countries. According to Saunders, "this power enables the commonwealth to control imports and exports and may be used with environmental objectives in mind" (Saunders 1996:63). Legislation evoking these powers includes action to prevent export of sand from Australia in 1975, the Ozone Protection Act (1989), the Wildlife Protection (Regulation of Exports and Imports) Act of 1982, and the Hazardous Waste (Regulation of Exports and Imports) Act of 1989 (Saunders 1996; Kellow 1996; Anton, Kohout, and Pain 1993), under which the movement of wastes is controlled by the use of ministerial permits (Lipman 1990:286). Thus, Australia is the *only* case here where waste import and export permits are issued not by state or local authorities but by central government officials. The commonwealth also controls the issuing of permits for interstate movement of wastes.

The implications of the above for waste importation, and indeed exportation, are significant: of all the cases examined here, Australia is the one that comes closest to a centralized waste management system with respect to the hazardous waste trade. In other words, government actors do not have to cope with the coordination and informational asymmetry problems identified in Britain.[30] In turn, this fits the prediction of the hypothesis regarding regulatory structure: The more centralized a country's structure of waste management, the less likely it is to be a waste importer.

Regulatory Style: Access to the Policy Process

With respect to the main indicators used to determine policy access, Australia has a relatively open system of environmental regulation and policymaking.[31] Environmental groups have benefited from commonwealth involvement in environmental decision-making and disputes. On the whole the government in Canberra has been responsive to the demands of the electorate and environmental organizations, in part for electoral reasons and in part because the government needed a legitimating base for expanding its activities in this area.

The Australian electoral system has militated against the emergence of a Green Party with significant representation at the national level. It does, however, allow voters with environmental concerns to either register a protest vote or vote for the Democrats, Australia's small third party, which has laid claim to the environmentalist banner.[32] This has meant that the main parties—in particular the Australian Labor Party, or ALP—have often taken on board environmental concerns in order to garner the Green vote.[33] Elections for the Australian House of Representatives are held using a preferential system. Voters rank a selection of candidates, and preferences are redistributed once first-choice candidates are eliminated, hence the scope for a protest vote. However, the practice of compulsory voting, as one analysis puts it, means "it is ultimately compulsory to cast a vote for a major party. Votes cast for minor parties are thus less likely to cost a candidate office as they might in a first-past-the-post system" (Doyle and Kellow 1995:129).[34] Australia's Senate elections, on the other hand, are run on a proportional basis, and candidates need only win a quota, or 7.7 percent of the total votes, to achieve representation (ibid.:132). It is here that the Democrats, and at least one Green Party member, have been able to gain and maintain a small but significant share of seats—on at least two occasions they have held the balance of power in the Senate. State governments, with the exception of the Tasmanian government, have been both less responsive to environmental claims and less fertile ground for Green electoral activity than the commonwealth government.[35] This has been due in part to differing electoral systems, and in part to their championship of development issues and interests.[36]

The environmental movement in Australia has therefore tended to target the commonwealth government in order to achieve its goals (Doyle and Kellow 1995:129).[37] Reforms introduced by successive governments have opened the environmental policy community to a broad spectrum of societal interests, including the key industrial and environmental groups. The environment ministry tends to champion its own interests and supporters, while industrial groups work through other ministries, many of which have established their own environmental agencies or working groups (Papadakis and Moore 1994:344). At the same time, the Labor government in the final years of its administration began to follow what has been termed a neocorporatist approach in its long-term environmental planning. This approach involved both pro-environmental and pro-development interests (industry and labor) in the consultative process over the ecologically sustainable development (ESD) process (Downes 1996).[38]

At the national level, then, Australia has a pattern of government-industry-society relations that is fairly balanced in terms of access to the policy process. At the state level it seems that industrial interests are more incorporated into decision-making processes; however, this is balanced by the greater priority given to environmental issues by the commonwealth government. Pakulski, Tranter, and Crook refer to the "routinization" of environmental concerns in Australia during the 1990s when they argue that environmental issues have become a fixture on the Australian political scene: not in the sense that "old politics have triumphed over new," but in the way environmental issues have taken their place in the array of concerns held by the average voter and "loosened up" traditional voting patterns (Pakulski, Tranter, and Crook 1998:250).[39]

Controversy over Corowa

In the late-1980s a battle over the siting of a waste incinerator had important implications for Australia's engagement in the waste trade. In 1987 the Victorian, NSW, and Commonwealth governments set up an independent task force to examine options for building a national waste disposal facility, much needed in the face of increased international controls on waste exports. In September 1990 the task force announced its recommendations: first, that a comprehensive waste management and minimization program be established for southeastern Australia (effectively, the entire country), and second, that a high-temperature incinerator and associated waste transportation network be constructed at Corowa on the Victoria-NSW border (McDonell 1991:11–12). The facility was to have been publicly owned but privately managed. This incident, and its aftermath, illustrates the dynamics of state-society relations as mediated by institutional factors in this issue area and provides some leverage in explaining Australia's risk-averse attitude toward the waste trade.

The outcry following the announcement of this decision was tremendous. While waste-generating industries and the waste disposal industry welcomed the construction of the facility, many key environmental groups, including Greenpeace and groups local to Corowa (although not the Australian Conservation Foundation, the largest environmental group), vehemently opposed it, as did the powerful local farming constituency. Public opinion polls revealed that voters had doubts over the feasibility of establishing a safe national transportation network to the plant (Anderson 1992b). The Corowa proposal also coincided with a couple of other large-scale, state-driven projects, including a fast rail link between Melbourne and Sydney that would have cut through several national parks.[40] By November 1990 the proposal was dead in the water, and the government was forced to set up another task force to address the issue. Corowa is not the only waste disposal site under contention. Throughout the early-1990s there was a long-running dispute over a landfill site on Coode Island—near some of the poorest parts of Melbourne—that had emitted waves of toxic gases over surrounding areas. The government initially suggested relocating the dump to a site located on a wetland protected under the Ramsar Convention, but after protests finally withdrew the proposal in 1997. Subsequent to 1991, roughly $A20 million have been spent on cleaning the Coode Island site.[41]

Australia is now moving toward a network of disposal sites, with an emphasis on developing new technologies. Research continues into the viability of alternative waste disposal methods, including more environmentally sound disposal techniques (some of which are now in operation on a small scale), waste minimization and recycling schemes, and portable waste disposal units (Anderson 1992a).[42]

McDonell ascribes the level of opposition to the waste incineration plant to differences between official and local/opposition assessment of the risks, costs, and benefits involved with the project, and to problems of trust and communication between government actors and the affected parties when facing large-scale policy change.[43] This assessment shares much in common with the situation in the other countries in this book; the difference is that opposing forces in Australia were able to voice their opinion in the policy process in ways that fundamentally affected the outcome. Existing channels of access to the policy process—the importance of electoral considerations and access to the immediate policy community—are crucial in understanding why efforts to set up a national waste management facility have failed, and hence why Australia is unable to effectively dispose of hazardous wastes.

Australia: Concluding Discussion
An institutionalist argument provides a powerful explanation for Australia's low waste importation propensity. Societal groups opposing the waste trade are clearly represented in the policy process, and the particular characteristics of the Australian electoral system ensure that governments try to be responsive to the environmental preferences of Australian voters. Australia's federal system is also important. Indeed, Australia provides the sole case examined here in any depth of a relatively centralized, and hence more easily coordinated and monitored, waste management system.

Australia has experienced some problems in its efforts to reduce reliance on hazardous waste exportation over the long term. In 1992 the government announced a two-year moratorium on exports to less developed countries and exports for final recycling (see table 6.1), despite the lack of domestic alternatives. This moratorium was extended into a permanent ban in December 1996 under the same conditions. In Basel Convention negotiations, Australia supports both the opening up of Annex VII membership to any countries that qualify and the allowance of Article 11 agreements (see chapter 2).[44] Between November 1997 and October 1998, Australia issued permits allowing waste export for recycling (primarily to Western Europe and the U.S.) totaling 82,493 tonnes; import permits in the same period totaled 4,145 tonnes. Unlike agencies in most countries, Australian authorities post full details about permit holders, waste totals, types of wastes, and waste destinations on the Internet.[45] However, a conclusion reached in 1990 rings at least partially true today: "Australia has achieved a great deal in the last few years but overall, has not kept pace with the rest of the industrialized world in developing regulatory arrangements, institutions and facilities for waste management" (Lipman 1990:291).

Two developments in Australian practices and policies are worth watching. First, the Liberal Party governments of the late-1990s have cut back government involvement in the economy. The implications of this change for environmental regulation have yet to be assessed. Second, Australian firms have announced their intention, following the principles of the Basel Convention, to begin exporting waste technologies to less developed countries. Which technologies are developed, and the extent to which technological transfer occurs, will be of much interest to those concerned with questions of hazardous waste management on a global scale.

Japan

Japan is a useful case for this analysis in several respects. First, it is a non-EU OECD country. More important, although it resembles Great Britain in many ways—including close government-industry relations and a diffuse regulatory structure—it is not a net importer of wastes.[46] In this section I examine why this is, and the extent to which such an outcome can be reconciled with the institutional model. After outlining the major features of Japan's system of environmental regulation, I focus in particular on two factors that differentiate the Japanese and British cases. First, the structure of the Japanese waste disposal industry is such that the importation of wastes is neither feasible nor desired. Industry structure in fact comes close to matching a model of perfect competition, and specialization (firms either collect or dispose of wastes, not both) makes importation extremely impractical.

Second, and more important, Japan was rocked by a series of environmental crises during the 1960s and 1970s—most notably, the identification of Minamata Disease. The institutional response to these crises was massive and, reinforced by a long political/bureaucratic memory (lengthy dominance of the Liberal Democratic Party; bureaucrats remaining in position), has continued to influence Japanese environmental policymaking and fostered a risk-averse attitude toward imported environmental problems. In the conclusion to this section I summarize the factors that might lead one to expect high levels of waste importation in Japan and the factors inhibiting such practices.

Early Development of Japan's System of Environmental Regulation

In Japan, perhaps more than any of the countries included in this analysis, the evolution of environmental policy from the early-1960s to the present has been shaped primarily by the changing dynamics of government-industry relations, particularly in terms of regulatory style. Owing to its extensive and extremely successful industrialization in the years following World War II, Japan has also been one of the countries where the effects of uncontrolled pollution were most rapid and severe (Tsum and Weidner 1989; Munton 1996b:183).

As with most industrialized countries, Japan's current system of environmental regulation dates back to a flurry of legislative activity in the late-1960s and early-1970s. During the 1950s and 1960s—the post World War II recovery process—environmental goals were nonexistent on the policy agenda. Pharr and Badaracco term this period one of tacit collusion between government officials and industry (Pharr and Badaracco 1988). It took the emergence into the public eye of four very serious pollution incidents, which resulted in a large number of deaths and permanent disabilities, to catalyze government action.[47]

A combination of community-based environmental activism, media attention, and foreign pressure led to the passing in 1967 of the Basic Law for Pollution Control. In its first version this law was heavily criticized, especially on the basis of the "harmony clause," which put economic prosperity on an equal footing with environmental protection (Weidner 1989a; Pharr and Badaracco 1988). Thus in 1970 the so-called Pollution Diet introduced fourteen new environmental laws into the statute books, ushering in a period of conflict between government and industry representatives. This conflict lasted until the mid-1970s, when a more cooperative relationship was reestablished (Pharr and Badaracco 1988). Whether this period of cooperation represents a victory for anti-environmental interests or a reorientation in the goals of the major firms remains open to debate. However, most agree that Japan has evolved a distinctive style of environmental regulation that differs in many significant ways from its OECD partners.

Structure of Hazardous Waste Management in Japan
Japan's high population density and high level of industrialization have made the management and disposal of industrial wastes a policy priority. However, the evolution of waste management policies in Japan has been complex and piecemeal; not until 1993 were laws treating hazardous waste separately from industrial waste enacted to govern the collection and release of data on its generation and disposal.[48]

Responsibilities for waste management in Japan are divided among different levels of government and government ministries. At the center, the Environment Agency and the Ministry for Health and Welfare are responsible for planning and coordinating environmental policies, establishing standards, and collecting and publishing data regarding wastes and waste disposal. Powers of policy implementation and waste disposal site control and licensing are assigned to the forty-seven Prefectures. The Environment Agency, established in 1971, is fairly weak in terms of its size, funding, and political clout. As many of its potential capabilities are in fact held by other ministries—including the powerful Ministry of International Trade and Industry (MITI) and the Ministry of Finance—or the local governments, its role has been primarily to act as liaison between the government and citizen groups (Barrett and Therivel 1991:13–14).

The prefectures, which generally have had more autonomy in the area of environmental policy than in other policy areas, "are authorized to establish more stringent regulations for air and water pollution control than the national level" (OECD 1993d:24, 58). Thus the severely polluted cities, Tokyo in particular, are noted for their activism in setting precedents for environmental policies. Unlike local authorities under Britain's unitary system, the prefectures have a high capacity, in terms of designated authority and resources, to implement environmental policies governing their resident householders and firms (Reed 1986).[49] "[L]ocal authorities get about 40 per cent of their revenue from the central government, so few of them can afford to directly oppose central government policy for fear of having these funds curtailed" (Barrett and Therivel 1991:15); however, they are less constrained by economic interests than is the central government, and they are considered to be responsive to citizens' and environmental groups (ibid.).

Relations between local and national politicians are fairly harmonious; Reed reports that, unlike in Britain, levels of partisanship are low (Reed 1986:42). However, "an elaborate consensus has to be worked out before environmental laws can be implemented, first between the EA, the other ministries and the LDP, and then between the local government, the polluter and the regional administrators," a factor that "complicates and delays the implementation of legislation and policymaking" (Barrett and Therivel 1991:73). In addition, many prefecture governments—in particular Tokyo and Yokohama during the 1960s and 1970s—have tended to be led by the opposition parties (who are more sympathetic to environmental concerns), thus further contributing to the complexity of center-local policymaking and implementation processes (Cole 1994:84). By setting precedents in pollution control policy, the prefecture governments challenged the national government to implement nationwide regulations, at the same time revitalizing local government in ways few expected (Reed 1986:46–51).

Japan's industrial waste disposal industry is largely in private hands, and many firms have onsite facilities. The Health and Welfare Ministry has allocated to each prefecture a Public Center for Waste Treatment "meant to ensure that there is sufficient environmentally sound capacity to treat waste (including special waste) in each prefecture" (OECD 1993d:59). As Beecher and Rappaport note, Japan makes use of public-private partnerships in constructing and running facilities (Beecher and Rappaport 1990:39). Disposal firms are typically small, and industry concentration is fairly low; this is in part due to the fact that over half of all industrial wastes are dealt with onsite by the generating firms. According to a 1996 report from the U.S. Department of Commerce International Trade Administration (Report # ISA 9053, June 29, 1996), as of 1992 Japan had 75,321 registered industrial waste treatment firms, of which 69,070 engaged only in collection and transportation of wastes. According to *Haznews*, only 6.6 percent of permits issued are to firms offering complete waste management service.[50] This separation of functions is another feature of the Japanese system. As might be expected from these figures, the average size of these firms is very small: only the top four or five maintain a workforce over 100; the rest average about 10–20 workers. Entry into the market by foreign firms is virtually nonexistent. Commercial disposal facilities rely primarily on incineration and increasingly scarce but secured landfill sites as the main routes of hazardous waste disposal; Japan's rate of waste incineration—50 percent of its toxic wastes are disposed of through high temperature incineration—is the highest in the world (Vogel 1992:266).[51]

Thus Japan's structure of waste management and regulation is best characterized as decentralized. As in Britain, there is a three-tiered allocation of responsibilities to the national government, local governments, and the waste disposal industry; moreover, environmental responsibilities are dispersed among different ministries at the national level. Unlike Britain, the level of conflict between national and local authorities in Japan is low, and authorities have fewer financial incentives to engage in waste importation or to abrogate regulation of waste importation. However, the potential for lack of coordination under this particular configuration is high and, as Barrett and Therivel (1991) point out, a significant obstacle to the development of comprehensive environmental policy in Japan.[52]

Regulatory Style: Access to the Policy Process

The Japanese policymaking process is noted for its reliance on a close and stable relationship among government, the bureaucracy, and industry—especially during the years of LDP (Liberal Democratic Party) dominance of the Diet—a relationship that excludes representatives of societal groups from the policy process (Barrett and Therivel 1991:8). This relationship is evidenced via informal means[53] and by the three groups' strong mutual commitment to the economic growth process and consensual rather than conflictual decision-making procedures, insulated from broader societal interests.

On the whole, environmental policy has not been an exception to this rule. The Environment Agency is charged with policymaking responsibilities, but it is heavily constrained by its junior role in relation to the big ministries, which have to consent before any bill drawn up by the EA can be presented to Parliament (Barrett and Therivel 1991:14). Thus, more than in the other cases examined so far, the environmental policy community in Japan is constrained by interministry relations and jurisdictional disputes, which tend to work against the efforts of the Environment Agency to implement fundamental policy change. Public opinion and environmental groups are excluded from this decision-making process: "Public participation is extremely limited: it is generally restricted to letters of complaint, opinion surveys and public hearings. Hearings, when they take place, tend to be strictly regulated explanatory sessions rather than forums for meaningful participation. Despite continued calls by citizens and lawyers for the introduction of a freedom of information act like that of the U.S., the national government is unlikely to expand public participation. Instead, it is promoting the sense that the administration can be entrusted to control pollution problems" (Barrett and Therivel 1991:75–76).[54]

The Japanese government has only recently begun to make data on hazardous waste generation and disposal available to the public. Reasons cited by the Ministry for Health and Welfare for withholding this information previously included "toxic waste disposal is not a problem; there are few cases of illegal dumping; and release of such information could cause public unease" (Delahunt and Burhenn 1991:61). In turn, this reflects the paternalistic tradition of state intervention in Japan, whereby the government is perceived as the best protector of citizen interests, a tradition broadly accepted by the public (McKean 1981; Cole 1994). Thus the environmental policy process in Japan is characterized by the fact that key decisions made by the immediate environmental policy community must also be approved by the main ministries, whose support of efforts designed to protect the environment is by no means given and who are also more likely to support a technocratic policy approach over a more "ecological" (or even inclusive) one.

This pattern of public exclusion from the policy process is mirrored in Japan's broader institutional structures. For example, the single non-transferable vote electoral system that prevailed until 1995 militated against the representation of smaller parties. Indeed, it was instrumental in perpetuating the dominance of the LDP (and the concomitant weakness of the opposition parties) until the early-1990s, when the LDP, rocked by a series of political scandals and corruption charges, was voted out of power.[55]

As a unitary rather than federal system of government, too, another set of access channels available to German citizens (for example, via the representation of the Länder at the national level) is closed to the Japanese public. On the other hand, Japanese prefectures have greater autonomy and capacity to set and implement environmental policies than British local authorities. Hence representatives of public groups have been involved in drawing up pollution control agreements between local governments and industry, including issuing permits to waste disposal firms. Increased involvement of local citizen groups has resulted in more stringent pollution control agreements in recent years, although the degree to which they are implemented is questionable (Barrett and Therivel 1991:78–79).

Access to the policy process via the court system is also limited. Disputes among policymakers are rarely submitted to such public scrutiny, and there have so far been few examples of preventive litigation. However, the judiciary in Japan has made an impact through cases where pollution victims have sought compensation for their injuries from industry, thus playing, as I argue below, an important, albeit indirect, role in determining the risk-averse nature of Japanese environmental policy and leading to Japan's reputation as "a world leader in redressing pollution problems (compensation for pollution victims) rather than pollution prevention" (Lehman 1992:725).

Regulatory Style: Mode of Policy Implementation

Japan's mode of policy implementation does not fit very neatly into the categories established in chapter 3. Government actors control industry through the use of "administrative guidance," a flexible tool that relies more on "suggestions, requests, encouragement or warnings" than on prosecution or sanctions for violation of regulations (Barrett and Therivel 1991:16). On the other hand, Japan employs strict emissions and environmental quality standards for the implementation of policy goals, and is noted for its technocratic approach in this regard (Weidner

1989a). Weidner summarizes the three main features of Japanese "technocratic activism": "In the development of specific pollution control measures, the emphasis is on their technical feasibility rather than on their legal foundations. For selected problem areas, stringent ends-means relations are set up, and, the goals to be reached are made very explicit, not intentionally vague.... To achieve environmental policy objectives, a short-term, and sometimes medium-term deadline is envisaged, and implementation is strictly supervised, say, by a comprehensive and sophisticated monitoring system" (Weidner 1989a:529–530).

Thus the Japanese mode of policy implementation is best characterized as mixed. It advantages industry to the extent that Japan's reliance on "technological fixes" for environmental issues more often than not involves an abrogation of control by political and social interests to economic ones, notably the firms themselves.[56] This flexibility is mitigated by the fact that firms are held to common standards, which in turn eases the burden of monitoring and reporting on their activities.

Effects of Japan's Regulatory System on Actor Behavior and Incentives

Government-Industry Relations and Environmental Policy As in Great Britain, the Japanese policy process is weighted in favor of industrial interests to the exclusion of societal interests. Government-industry relations in Japan have always been something of a loaded topic for many academics, and many well-known discussions of this subject border on caricature.[57] The Japanese state is certainly more powerful and interventionist than its British equivalent.[58] However, the relationship among the party (parties) in power, the bureaucracy (represented by the major ministries), and industry (the key firms and their peak associations—the *keidanren*) is one of interdependence, with a strong sense of common goals and interests (Boyd 1987).[59] Boyd argues that the links between government and industry, which "facilitate the movement of ideas, [and] the formation and representation of interests ... are not remarkable: many similar channels exist in other industrial nations. What is distinctive is the extensive use made of them. This is a consequence of the insulation of the industrial policy-making and implementation process from public debate" (Boyd 1987:65).

Apart from a brief period in the early-1970s, there have been few periods where government and industry preferences regarding the environment have diverged enough to determine which, if either, ultimately has the upper hand. For example, during the 1973 energy crisis and its aftermath, Japan's unified response across all sectors of society seems remarkable from the vantage point of more pluralist political systems. Pharr and Badaracco (1988), who argue that Japanese environmental policy has to a large degree been shaped by a changing government-industry relationship (from collusion to a brief period of conflict to cooperation), identify the main period of government-industry conflict as lasting from the mid-1960s to the early/mid-1970s. They argue that this period represented the main phase of environmental activism on the part of the government, which was able to push through the main pollution laws (as described above) over the objections of the main firms. Yet this period—characterized also by unparalleled levels of activism by local governments and citizen groups—proved to be both unique and of short duration; since then a more cooperative relationship has prevailed. This shift was occasioned by, among other factors, the post-oil shock recession, the resolution of the four pollution suits, and MITI's wresting from the Environment Agency control over certain emissions regulations. (Pharr and Badaracco 1988:243–251).

Government-Society Relations: The Japanese Environmental Movement
In contrast to industrial interests, the environmental movement in Japan has, especially in recent years, been fairly weak, and dissipated in terms of its influence. The focal point for environmental organizing in Japan is the "citizens' movement," of which, according to one estimate, there were roughly 20,000 in 1991 (Crump 1996:115). Citizens' movements generally arise at the local level, mobilized by a single issue. Groups join together to combat this issue at the national level, and their tactics tend to be highly confrontational—as demonstrated by the often violent protests, which peaked first in 1968 and then again in 1989, over the building of Narita Airport (Barrett and Therivel 1991:19).

Yet there has been little sign of effective translation of citizens' movement demands into national policy, nor has a cohesive, nationally-based, and well-funded environmental movement—along the lines of a

Sierra Club or a German Green Party—emerged. More often than not, citizens' movements remain local. Most analysts trace this to the relative impermeability of the Japanese policy process and to the government's tactic of providing compensation for pollution victims rather than taking anticipatory action in the face of environmental harm.[60] Others relate it to a public perception of pollution that questions the actions of individual firms but not the broad concept of development itself (Munton 1996b:192).

Public interest in environmental degradation seems to have peaked in the mid- to late-1970s, in the aftermath of the big four pollution lawsuits. Since then, environmental issues have declined somewhat in salience as media interest has moved on to other matters (Cole 1994; Barrett and Therivel 1991). This trend has been related to a number of factors, including a widespread perception that environmental problems are being adequately addressed, a resurgence of concern with economic performance, and a feeling of lack of efficacy of environmental activism (Schreurs 1997; Vogel 1992).

Japan as Nonimporter of Wastes

The above analysis suggests that Japan has a regulatory climate extremely conducive to waste importation. Societal interests opposing waste importation are excluded from the policy process. In the absence of direct evidence on government/bureaucracy actor preferences regarding waste importation, it would appear that should industry wish to engage in waste importation, it would find few institutional obstacles in its way. Furthermore, Japan's regulatory structure, with its decentralized responsibilities, privatized disposal industry, and a very marginalized Environment Agency, suggests that the sorts of principal-agent problems plaguing the British system could also exist in this case.

However, Japan is not a net importer of hazardous wastes: its overall environmental policy stance in this area is best characterized as risk-averse. As table 6.1 demonstrates, Japan in fact comes close to self-sufficiency with respect to the waste trade.[61] In the following section I argue that this stance can be related to two factors: the structure of the waste disposal industry in Japan and, most importantly, the legacy of the "Big Four" pollution cases of the 1970s—a legacy amplified by the long institutional memory of Japan's political system. The analysis leads to two conclusions: first, that industry preferences in Japan are, as a result of both the above factors, decidedly not proimportation; second, even if this were the case, existing institutional and ideational obstacles would make it much less likely for government actors to allow waste importation.

Japan's Waste Disposal Industry As the above discussion of the structure of the waste industry in Japan demonstrated, there is little evidence that the waste disposal industry per se is a powerful actor in the policy sphere, especially as the larger firms tend to dispose of their wastes onsite. Instead, the most important interactions in the environmental policy arena occur between government and the large industrial production firms and conglomerates; hence, environmental debate is focused much more on the activities of pollutant and waste generating firms than of disposal companies (Pharr and Badaracco 1988).

The most convincing explanation of the structure of Japan's waste disposal industry lies in the separation of disposal, transportation, and collection activities and in the very small average size of its firms. These factors, a function of the licensing and permitting procedures of local governments, militate against the feasibility of waste importation from abroad. In addition, and related to the next set of arguments, Japan's highly protectionist stance with respect to the activities of foreign firms has precluded the entry of foreign firms into the Japanese market (Waste Management International made a brief foray in 1991, but soon withdrew).

The "Big Four" Pollution Cases and Risk Assessment in Japan During the 1950s and 1960s, it became increasingly apparent that diseases related to pollution and industrial waste were devastating many communities. Four cases in particular—mercury poisoning in the Minamata and Niigata areas, a collection of pollution-related diseases in Yokkaichi, and cadmium poisoning ("Itai-Itai" disease) along the Jintsu River—catalyzed judicial and legislative action when scientific investigation formally established causal links between the activities of certain firms and the incidence of disease.[62]

In the early-1970s the courts, despite their previous lack of environmental activism, ordered that the firms responsible for the environmental damage pay out huge sums in compensation to the victims. These decisions led to a system of victim compensation unparalleled in the rest of the world and set important precedents for subsequent cases, including shifting the burden of proof onto firms, allowing for statistical connections in the absence of direct causation, and allowing plaintiffs to sue individual firms rather than entire industries in the case of damage (Crump 1996:21). Such cases have continued to arise, not only at the national level but also at the level of local government—where agreements have often gone beyond basic legal requirements (Weidner 1989a: 490)—although restrictions were placed on compensation awards in 1988 official rulings.

The implications of these precedents for the large corporations that make up a substantial part of Japan's industrial infrastructure are obvious: "The vastly increased risk that companies ... face of having to pay compensation was a great stimulus not only to undertake remedial action but also to adopt preventive environmental measures" (Weidner 1989a:538). Companies became less and less willing to undertake activities that would be seen to have a direct and negative effect on the environment and human health. This has been a major deterrent for waste importation, affecting the incentives facing both firms and government actors that might otherwise wish to engage in or allow waste importation.

Weidner and others also argue that the threat of endless litigation catalyzed the very technocratic style of environmental policymaking subsequently displayed in the 1970s.[63] Compounding this, Japan's very stable governance structures—characterized, for example, by the long tenure of key office holders in both the bureaucracy and government—has led to a long "institutional memory." Thus events that in other, more transitory, political systems might have faded into history continue to cast a long shadow on current policy, although as some argue, this shadow's influence has in recent years begun to fade (Crump 1996; Pharr and Badaracco 1988).[64]

Vogel (1992) discusses the factors that have influenced a highly risk-averse attitude on the part of Japanese government and society with respect to foreign goods, services, and indeed pollutants. Although the term "risk" is avoided in policy discourse, he identifies a clear set of indicators that suggest official policies in the realms of consumer and environmental protection in are governed by a highly risk-averse attitude. This phenomenon he relates in part to Japan's protectionist impulses and in part to differing classifications of safe and unsafe pesticides and drugs. Antinuclear sentiments, a legacy of the bombs dropped on Hiroshima and Nagasaki at the end of World War II, also plays into Japan's aversion to imported risk. Vogel's analysis has interesting implications for waste importation, for it suggests that while Japan is a noted laggard in negotiations over global environmental issues and appears to be less concerned than its Western counterparts about domestic sources of industrial pollution, it views potential or real pollutants from abroad with the highest possible degree of suspicion: "Many Japanese appear to regard foreign products and technologies as inherently less safe than ones produced in Japan: they trust neither foreign manufacturers nor foreign regulatory authorities" (Vogel 1992:153; see also pp. 148–149).

Japan: Concluding Discussion
Japan presents an anomaly to the institutional model of legal hazardous waste importation patterns: despite the strong resemblance of its regulatory style and structure to Britain's, it in fact does not import hazardous wastes. Instead, Japan's tendency towards autarky in waste importation is related to external factors not present in other cases. The most important of these was a clear period of environmental crisis faced by Japanese society throughout the 1960s and 1970s as a result of industrial pollution. The "resolution" of this crisis, primarily through court orders for firms to pay compensation to victims, in turn has fostered an unwillingness to engage in practices seen as particularly risky to public health. Following Vogel's argument about Japanese risk attitudes, this would apply in particular to pollutants, such as wastes, imported from abroad. This is not to say that institutional factors are not important. For example, one of the reasons these cases have remained influential in policymaking is that the long spell of LDP dominance and the lengthy job tenure of key bureaucrats and industry leaders have kept the memories of that time and its results fresh.

Recent analyses of Japanese environmental policy have pointed toward some worrying domestic trends, even as Japan is beginning to address its international reputation in the environmental sphere (Crump 1996).[65] For one, the tradeoff between economic growth and environmental protection is viewed in starker terms in Japan. As economic recession has set in, economic concerns have begun again to dominate environmental concerns. Some—Helmut Weidner, for example—argue that the very technocratic nature of Japanese environmental policy militates against the development of a more preventive approach, as would be embodied in the adoption of the Environmental Impact Assessment Bill.[66] Whether the recent political upheavals and reform imply "business as usual" for the environment or more fundamental change is still an open question. Problems of waste management in Japan, as with the other industrialized nations, remain paramount, and in the absence of direct action are only likely to increase. Many analysts believe Japan's biggest challenge is the need to develop both a preventive orientation in combating environmental degradation and effective resource management in the face of projected scarcity.

Conclusion

The results presented in this chapter on the whole support the thesis that certain features of countries' systems of environmental regulation and political institutions are determining factors of their waste importation propensities. In more general terms they demonstrate that the model developed in chapter 3 can be extended to other OECD countries. Table 6.2 shows a clear correlation in the French and Australian cases between their political structures and styles of environmental regulation and those of Britain and Germany, respectively—a correlation that carries over to their waste importation propensities. In each case I demonstrated how preferences of different groups are mediated by these two factors. In France, groups opposing waste importation have been filtered out of the policymaking process, while in Australia societal opposition was able to block the construction of a high-temperature waste incineration facility designed to reduce Australia's dependence on waste exportation.

Japan presents an exception to the general argument: despite a regulatory system that corresponds to Britain's in many respects, it is not a waste importer. This is due primarily to the continued effect of factors exogenous to the model, namely the "Big Four" pollution cases that rocked Japanese society through the 1950s and 1960s and into the 1970s, when the cases were finally settled. These fostered a risk-averse attitude on the part of industry and government actors—the former for fear of having to pay out large amounts in compensation, and the latter for fear of societal opposition and unrest. Although the continued effect of the "Big Four" cases can be traced to institutional factors (e.g., a political system that fosters long tenure and continuity in government and administrative posts), these are not the same institutional factors on which this argument is built.

Beyond the empirical results, the French, Australian, and Japanese cases also raise interesting issues for the theoretical approach. First, the hypothesis regarding mode of policy implementation, at least in these three cases, is less important than those concerning structure and access. In Australia and Japan it was hard to identify mode of implementation as an independent feature, and it seems it does not play an independent role in these cases. Second, these interpretations of national environmental policy practices and styles continue to run counter to conventional interpretations of public policy in these countries. This has to do with the nature of environmental degradation as a policy issue. Not only do its contours change rapidly over time with the advent of new problems and new understandings about them, but it also has effects at all levels: local, national, and international. The countries that can best pay attention to these dynamics tend to be those with more coordinated and open regulatory systems.

7
Conclusion

This project set out to establish whether key procedural differences between national systems of environmental regulation can explain why some countries, industrialized democracies in every case, should—in the absence of coercion—willingly take on the risks of disposing of hazardous wastes that other countries do not want. The waste trade is a highly risky activity, associated strongly with environmental damage, market failures, and issues of equity and social justice (the predominant theme in the literature on the trade so far). The movement of wastes across national frontiers is agency-driven: deliberate decisions are made by waste disposal firms to accept foreign wastes for disposal and by the government agencies that issue waste importation permits or otherwise allow hazardous wastes to enter the country. At the same time, waste importation, along with many other issues concerning hazardous waste management, is often strongly opposed by a wide range of forces, usually including environmental groups and the weight of public opinion.

Despite these risks, empirical examination reveals a wide array of different behavior across OECD countries: some are large net importers of wastes while others are net exporters, and there are a variety of practices in between. Britain and France are the world's two largest net importers of hazardous wastes, positions they have maintained since waste trade records began in the mid-1980s. Germany and Australia, on the other hand, are both exporters: Germany is probably the world's largest waste exporter, Australia much less so. Finally, Japan is the country that comes closest to autarky with respect to the waste trade. These differences hold despite broadly similar preference configurations among the various stakeholders in the waste trade across these countries, and despite the absence of significant differences between their relative economic positions or their membership and/or stated support for the relevant international agreements and regimes.

I begin this chapter by summarizing the main insights and results of this study and by demonstrating why an explanation of waste trade tendencies based on regulatory differences is superior to alternatives. I continue by examining the broader implications of this argument for studies of international and comparative environmental policies. Next I push the conclusions beyond the comparative into the dynamic, examining how regulatory changes at the national level, especially those originating from international or transnational sources, are dealt with in different approaches, and I consider how an institutional approach that recognizes the role of agency helps in interpreting institutional change in the context of European integration. I then address the likely evolution of the international waste trade regime and the prospects for banning the trade. I also address how a changing public-private balance worldwide in hazardous waste management is affecting the trade and note some of the lessons industrialized countries are showing evidence of learning. Finally, I discuss some policy prescriptions arising from this work for the developed countries, being careful to bear in mind the difference between *causal* variables—of most interest to social scientists—and *malleable* variables—of more interest to the practitioner. Unfortunately these are not always one and the same; the hard part is to work on the latter in the context of wider, less malleable, but nonetheless important variables.

Results: A Summation

Contending Explanations

In this analysis I have identified three contending explanations to account for differences in waste trade practices: the first based on rational cost-benefit calculations by state actors—the "economic nationalist," or "financial incentives," explanation (based on Montgomery 1992); the second based on comparative levels of disposal technology; and the third based on institutional factors—differences between national styles and structures of environmental regulation. Table 7.1 lists the arguments tested for each.

Table 7.1
Contending explanations

Economic-nationalist explanation
• The legal transfrontier movement of wastes (in either direction) depends on how government actors calculate the costs (or risks) and benefits associated with the outcome of this decision.

Comparative advantage explanation
• Countries with a higher spare disposal capacity are more likely to import hazardous wastes.
• Countries with more advanced disposal facilities are more likely to be involved in the legal importation of hazardous wastes.

Regulatory explanation
• The more decentralized a country's structure of hazardous waste management and regulation, the more likely it is to import hazardous wastes.
• The more closed a country's system of environmental regulation, in terms of allowing access to the policy process to a wide range of groups and interests, the more likely it is to import hazardous wastes.
• The more flexible a country's mode of policy implementation, the more likely it is to import hazardous wastes.

The economic-nationalist explanation proved problematic on a number of grounds, as demonstrated in the context of the British case. Calculating and quantifying the relevant costs and benefits and how they are distributed among different actors is almost impossible, given the vast differences between private and social assessments of associated risk. Hence, while there are certainly large financial gains to be reaped from importing hazardous wastes, it is uncertain whether these outweigh the social costs of waste importation, costs that eventually fall on government shoulders.[1] More basically, this argument does not stand up to comparative analysis on its own terms: it cannot explain why different countries—Britain and Germany, for instance—make such different cost-benefit calculations on the basis of ostensibly similar levels of risk. In fact Germany, with its higher levels of disposal technology and higher potential profits from providing waste disposal services, should be more inclined to allow waste importation than Britain, which has vastly inferior—and hence cheaper—disposal facilities. However, this is not the case.

The second part of the economic-nationalist argument—that rational, unitary state actors are able to arrive at these decisions autonomously of social pressures—is also problematic. In most cases, waste importation decisions are not made by a single official actor; rather, they are often the result of input by multiple regulatory agencies and/or levels of government, agencies that may or may not be subject to pressures from many different societal actors.

The comparative advantage argument also proved to be of limited explanatory power. It posits that advanced industrialized democracies import hazardous wastes on the grounds that they have both adequate spare capacity and superior disposal facilities capable of handling extremely toxic substances more safely and speedily than the facilities of their exporting partners. Across the board, arguments about spare capacity do not hold: in every case—with the possible exception of Germany, where disposal authorities are beginning to report that many facilities are indeed underutilized—government agencies and environmental groups report that hazardous waste disposal capacities are under extreme pressure.[2] All OECD countries are having severe problems siting and constructing new facilities, due to both societal and economic constraints.

The technology argument required further empirical examination. Results showed that in the two importing countries—Britain and France—disposal facilities are in fact alarmingly inferior. This observation held most strongly in the British case, where the main disposal route for hazardous wastes and residues is landfill, and where incineration facilities are considered inferior to those in many other Western European countries. The low disposal costs associated with these routes make these countries attractive destinations for waste exporters. However, they cannot explain why the authorities in these countries continue to allow waste imports, given the level of risk imposed on local populations and the high probability of substantial cleanup costs in the not too distant future. Germany, on the other hand, offers relatively high levels of disposal technology, yet authorities there (despite some evidence of industry support for this idea) show little willingness to import wastes from countries less able to deal effectively with them. Australia, finally, is something of a special case: it almost entirely lacks adequate facilities for dealing with hazardous wastes. Although such facilities have been proposed at various times, high levels of societal opposition influenced federal and state governments to halt their development and construction.

Institutionalist Explanation: Regulatory Structures and Styles
The effectiveness of societal opposition in Australia lends additional support to the most powerful explanatory account: that differences in waste importation practices can be traced to differences in national styles and structures of environmental regulation, and that in turn these differences at the international level help explain certain patterns of waste trading among OECD countries. Two features of countries' systems of environmental regulation determine waste importation patterns: their regulatory *structure* (the allocation of waste management and regulation responsibilities among agencies and levels of government and the structure and ownership patterns of the waste disposal industry) and their regulatory *style* (how state-society relations are played out in policy formation and implementation). This study identified two indicators of regulatory style. First, *access* to the policy process can be achieved via inclusion in the environmental policy community or through access channels afforded by a country's broader political opportunity structure. Second, a country's mode of policy *implementation* depends on whether policy standards and goals are implemented on a rigid, across-the-board basis or on a more flexible (ad hoc), case-by-case basis.

This is an institutionalist explanation; hence, the empirical analysis needed to demonstrate how these features constrained and/or facilitated the behavior of the relevant actors in such a way as to determine national waste importation propensities. The hypotheses discussed in chapter 3 generated a twofold argument. First, certain types of regulatory structure—notably, the highly diffuse structure exhibited by Britain —vastly complicate the monitoring problems facing policymakers at the center seeking to control, or at least coordinate, waste importation practices. Second, these regulatory features determine certain patterns of access to the policy process. In some cases, they filter out broader societal preferences in favor of industrial interests. In others—notably, where waste importation does not occur—they enable a wide range of actors to influence policy decisions and outcomes. Overall, these three

Table 7.2
Main variables and findings

	Great Britain	Germany	France	Australia	Japan
Structure of waste management and regulation	Diffuse	Coordinate	Diffuse	Centralized	Decentralized
Access to policy process	Closed	Open	Closed	Open	Closed
Mode of policy implementation	Flexible	Rigid	Rigid	Rigid	Flexible
Waste importation propensity	High	Low	High	Low	Low

Notes:
Structure of waste management and regulation = allocation of regulatory authority between national and subnational authorities, number of regulatory authorities, and structure of waste disposal industry (public or private; degree of competition).
Access to policy process = composition of environmental policy community plus a country's political opportunity structure (e.g., electoral system, judicial review).
Mode of policy implementation = use of standards; monitoring and enforcement mechanisms and principles.
Waste importation propensity = a country's status as net importer of wastes, measured over time.

factors—regulatory structure, access, and mode of implementation—combine to determine whether a country has a regulatory climate permissive of the importation of hazardous wastes by (usually) private actors or, alternatively, whether a country exhibits risk-acceptant or risk-averse behavior with respect to waste importation.

The results across cases are summarized in table 7.2. The two primary case studies were Britain and Germany. Britain fits the profile of a waste importer. A highly competitive, privately held waste disposal industry dominates a policy process that tends to exclude the views of the general public and environmental groups. This process of exclusion —which occurs not only within the "corridors of Whitehall" (the environmental policy community) but also through Britain's first-past-the-post electoral system and lack of independent processes of judicial review—reflects a long-standing tradition of cooperation between industry and government actors in the formation of public policy (Vogel 1986). Furthermore, powers of waste management are dispersed among many local regulatory authorities, which themselves have very little capacity to monitor the activities of firms under their jurisdiction, even though they, and not central government, have the authority to issue waste importation permits. There are few clear channels of communication between central government and local authorities, and their relationship has traditionally been highly conflictual. These problems have complicated the waste importation issue to the extent that the Westminster authorities have been unable to implement a ban they announced on waste importation in 1991. This factor is compounded by a mode of policy implementation that gives industry a large amount of discretion in setting and meeting environmental goals and standards. Although, as shown in chapter 4, local regulatory authorities are beginning to consider alternative strategies for dealing with this problem, progress toward this goal remains slow.

Germany's system of environmental regulation contrasts with Britain's in almost every respect. First, relationships among the various constituent units of its regulatory system are much more structured, and the structure of waste management in the Federal Republic is much more centralized. There are fewer regulatory authorities in charge of waste management (these are located at the Land level), and the waste disposal industry is either publicly owned and run or subject to high levels of government intervention and scrutiny. These factors alone mitigate the sorts of principal-agent problems that plague the British system. There are also many more points of access to the policy process in Germany. The electoral system, for example, has allowed the German Greens access to the national parliament, thus forcing the more established parties to take on board environmental concerns. There is also evidence that the policy community is opening up to include not only industry and scientific experts (the latter a category excluded in Britain), but also a wider range of societal interests. Finally, Germany has a famously rigid policy implementation style, based on risk-averse calculations of polluting effects (the "precautionary principle"), which gives industry very little leeway in determining its actions. The high level of legitimacy accorded to the German mode of state intervention in the actions of its constituent firms has also allowed Germany to develop and implement an extensive program of industrial waste minimization, embodied in the *Kreislaufwirtschaft* (Closed Circle Economy) ordinance.

Thus empirical analysis of Britain and Germany's environmental policies bears out the hypotheses developed about the effects of regulatory differences on waste importation propensities. In chapter 6 I examined France, Australia, and Japan in order to show how this model can be applied to other OECD countries. The first two cases again bore out the regulatory explanation. The most important factors in each were regulatory structure and access to the policy process. In France, for example, the ambiguities surrounding the delegation of environmental responsibilities to the relatively new regional tier of government has made it much harder for government actors to control the actions of importing firms. The closure of the policy process to interests opposing waste importation also contributes to the problem. The French government, too, is now following the British government's lead in developing a package of fiscal incentives to help stem the influx of imported wastes and to discourage the use of landfill as a primary mode of disposal.

Australia has the most centralized waste management structure of the cases examined here. Even though it has a federal system of government, the industrial activities generating hazardous wastes are concentrated in two states, Victoria and New South Wales, and most existing waste storage facilities are in Sydney, the capital of New South Wales. The most interesting feature of this case concerned the debate surrounding the construction of a national facility for waste incineration. The effectiveness of opposition to this plant demonstrates Australia's relatively open access system. Australia, too, of the cases here, is currently giving the most consideration to the development of "alternative" —generally, more environmentally sound—modes of waste disposal.

Finally, Japanese waste importation practices represented an anomaly to this model. While the Japanese regulatory system mirrors that of Britain in many respects, Japan it is not a net importer of hazardous wastes. This result can be related primarily to the continuing impact of several pollution scandals that came to light in the late-1950s and early-1960s, and whose effects in terms of victim outrage and compensation cases continue to resonate today, fostering a risk-averse attitude to importing hazardous substances. The long shadow of these events can be ascribed to institutional factors, but to different ones than identified here. More specifically, analysis showed that the structure of the Japanese waste management industry—one of nearly perfect competition—is not at all conducive to waste importation. Separate firms carry out waste collection and disposal activities, and there are no large waste disposal firms in Japan of the sort that import wastes in Britain and France.

In sum, the regulatory argument works well as an explanation of why some countries import more hazardous wastes than others. Not only does it perform better than alternative accounts, but it is also applicable across other OECD countries not included in this study. The regulatory systems examined in the case studies have been relatively durable over time; pressures for change have targeted perceived national needs: structure in Britain, waste minimization in Germany, and technology in Australia. The relative durability of these regulatory systems strengthens the argument in favor of the existence of institutional constraints on national actors. As data currently limited in scope becomes more readily available, and as more progress is made toward monitoring waste generation and disposal trends and harmonizing national definitions, extension of this argument to other cases will become possible.

Theoretical Implications: Generalizing the Argument

International Environmental Politics

There are many links between domestic and international political processes with respect to global environmental issues, yet to a great extent they remain underexplored in much of the literature so far. Domestic regulatory policies are, as many authors have pointed out (e.g., Mitchell 1994), vital for the effective implementation of international environmental agreements and indeed for other sorts of international regulatory agreements. Hence detailed analysis of the different national responses to questions of environmental degradation yields much that is useful for understanding these processes. Chapter 1 raised two issues that the field has not yet adequately addressed: the political factors that underlie global environmental degradation and the definition and measurement of regime effectiveness, both of which can be addressed within a domestic regulatory framework.

To understand how states respond to (and how they might contribute to) environmental degradation, both local and international, it is often necessary to look beyond questions of regulatory capacity or effectiveness. The countries examined in this study are all roughly comparable in terms of their overall levels of success in tackling key environmental problems. The important differences between them are procedural. Differences in the ways states deal with environmental issues can be seen at all stages of the policy process, as the above summary of results demonstrates.

Other important institutional differences can be derived from the five case studies. For example, interagency relations—part of a country's regulatory structure—matter. In Britain, France, and Japan, the environment ministry is marginalized in the political process; this is far less the case in Australia and Germany.[3] A country's tendency to frame waste management as a local rather than international issue also determines how it views the key problems involved. Finally, differences in modes of implementation help determine the amount of leeway afforded to industry in their actions regarding environmental protection.

Different national responses to environmental issues are brought to bear in international negotiations and in subsequent implementation of resulting agreements. At the Basel Convention negotiations, the OECD countries differed notably in their approaches to wastes destined for final disposal versus those destined for recovery, and countries with the most powerful domestic waste industry lobbies were much less in favor of imposing a ban on waste trading. More importantly, regulatory differences can have an important impact on both how and how well countries implement international environmental commitments—for example, Britain's pledge to end hazardous waste importation. Therefore an understanding of how different national modes and styles of regulation work helps us to predict the likely effectiveness of a given agreement (see discussion of the Basel Convention, below). It also addresses, in a way that goes beyond merely assessing state compliance, the issue of how to conceptualize or measure regime effectiveness in the absence of data on environmental changes that result from an agreement.⁴ For example, if there is little adaptability in either the agreement (to national differences) or the national regulatory system (to the terms of the agreement), then it is unlikely that the agreement will be effective. Alternatively, if the agreement imposes conditions that are impossible for a state to carry out given the nature of its regulatory system, again the agreement is unlikely to be effective.

Generalizing this sort of midrange theoretical approach to other international environmental issues and agreements allows an understanding of how countries implement the terms of an agreement and how governments seek to control the actions of the private individuals and firms under their jurisdiction. In chapter 2, I identified several key characteristics of hazardous waste management as an international issue. It is a local-cumulative issue with important transboundary spillover effects, and there is a high degree of agency in the transmission of the problem across national frontiers: a small group of relatively identifiable actors make decisions to ship wastes to other countries. Examples of highly comparable issues are the international timber trade, nuclear waste disposal, and illicit CFC smuggling.⁵ Nuclear issues have been especially volatile in the late-1990s in all of the case study countries, as they grapple with nuclear waste transportation and management issues that in the advent of several high profile cases have become international (O'Neill 1999b).

Comparative Environmental Policy
The theoretical approach of this book is derived from and extends the tradition of institutionalist analysis in the study of comparative public policy. Several broad implications of this work can be identified. First, the empirical work and methodological approach emphasize the extent to which hazardous waste management practices are embedded not only in wider rules and practices of environmental regulation, but also in much broader institutional structures such as a country's electoral system and polity type. The framework developed in this work can be relatively easily applied across a wide range of countries, despite the differences between them.

Second, polity structure—whether a state is federal or unitary—has a significant impact on waste trade and management practices in ways that are often counterintuitive. This is a complex issue area, with local, national, and international ramifications. Federal systems—like those in the U.S., Germany, and Australia—generally tend to have clearly defined and accepted divisions of power and responsibilities among different levels of governments. This means that they are much better than unitary states at overcoming some of the monitoring and coordination problems inherent in coping with complex, multilevel, and environmentally risky activities. In Britain and France, environmental powers tend to be devolved to poorly resourced local tiers of government, whose legitimacy (or social base) is correspondingly weak. It is likely that the insights derived from this study of the waste trade could also be applied to other policy areas displaying similar domestic/international linkages.

Third, I have demonstrated how regulatory systems interact at the international level to determine certain patterns of waste trading. Although many works in this field have compared different responses at the domestic level to international environmental issues (e.g., Boehmer-Christiansen and Skea 1991), few have examined the interplay between these national systems—for example, how Germany and Britain have become waste trading partners. In turn, these observations intersect with the literature on the links between international trade and the environment by showing how national regulatory differences can, in the case of harmful goods such as hazardous wastes, actually facilitate the movement of such goods across national frontiers.

Finally, this work overcomes problems inherent in many studies comparing the effectiveness of environmental policy across countries. Effectiveness is a slippery variable to measure and compare, unlike waste importation or exportation, where the data is easier to read and outcomes easier to classify. Many studies (Bernauer 1995; Vogel 1986; and Victor, Raustiala, and Skolnikoff 1998) discuss this problem at length. However, using net imports or exports of wastes as a proxy variable makes it possible to identify some measure of a country's risk propensity with respect to waste management, and makes possible some very interesting observations that would elude a more traditional study of regulatory success or failure.

Waste import or export figures indicate whether a country is risk averse (e.g., Germany and Australia) or risk acceptant with respect to waste management. Some questions remain. For example, are Britain's attitudes toward waste importation mirrored in attitudes toward other issues, or are there in fact significant differences across issue areas? Early analysis would suggest that the former is the case: Britain and France also host a large part of the world's nuclear reprocessing industry, for example. Also, while Germany's refusal to import wastes demonstrates a high degree of risk aversity in protecting its domestic environment, its willingness to export wastes abroad displays a comparative lack of concern for the environmental health of other countries. Hence, very quickly these questions run into issues of state sovereignty and interstate interaction, and with the issue of why some countries are more actively engaged in international environmental cooperation than are others.

Regulatory Change and European Integration

The theoretical approach of this book and the empirical cases presented intersect in many ways with debates over the direction, pace, and mode of regulatory change and with debates over European integration. Naturally, these debates are not mutually exclusive. The context in which national regulatory policy is made has changed so dramatically over the last decade that some question the utility of talking about national policy at all. National borders are becoming more porous, transnational communications among individuals and groups are infinitely faster and more frequent, and policies are being made with a greater understanding of the extent of global ecological interdependence. These days it really is possible to talk of an international community and of transnational networks (Keck and Sikkink 1998). Domestic contexts are changing too, with private and nongovernmental actors taking or being given a more extensive role in governing their own actions with respect to the environment. In three of the cases—Britain, Germany, and France—the EU has taken on an increasingly important role in issuing directives and policy guidelines then implemented by member state governments. At the same time, most of the countries examined here are in the process of overhauling their existing environmental regulatory systems, with perhaps the most extensive changes being undertaken in Britain, where administrative structures are being reorganized toward a system of integrated pollution control (IPC).

For scholars of environmental policy, the field is therefore opening up to the recognition of a variety of factors that enable or impede desired policy outputs and regulatory change. This book has studied key institutional parameters and actors in the context of a nested model of regulation: to understand hazardous waste politics, for example, it is also necessary to understand national politics. It is the interaction of these national systems that creates specific patterns of waste trade, links certain important actors (e.g., firms and NGOs), and helps generate the shape of international governance regimes. In order to simplify the theoretical analysis of this complex interaction, I chose to exclude explicit consideration of pressures for change originating outside the national systems. At the same time, this approach and some of the results found at national and local levels have important implications for the preexisting, rather tangled set of approaches to both regulatory change in general and environmental regulatory change as a result of European integration in particular.

This debate is indeed a complex knot of theories, approaches, and frameworks. Many are centered around the evolution of the EU, and most emphasize the importance of international and transnational driving forces of change. The empirical work presented in chapters 4, 5, and 6 clearly illustrates the continued importance of national differences in environmental policy and supports "bottom-up" perspectives on EU integration. However, close examination shows that policymakers are coming to share a certain set of ideas about the relationship between economic growth and environmental protection. These ideas, transmitted from the international level and from other states and organizations, have been (or will be) translated into national practice in ways that are mediated by existing institutional configurations.

In this section I focus on some approaches to understanding regulatory change—especially within the context of the European Union—and how well they capture both the dynamics of and differences across countries. I begin with a large-scale, structural approach: regulatory convergence. This approach is structural in that its causal impetus relies on broad, often global, macroeconomic change rather than on the actions or decisions of specific individuals or groups. Actors (nations, industries, and groups) react to these trends but rarely shape them. I then discuss more actor-based approaches that specifically seek to explain the evolution of EU environmental politics in the 1990s. These fall into two categories: "bottom-up" approaches, which examine the way member states shape transnational politics in different ways, and conflict/stalemate approaches, which ask why we have seen less progress than expected in this arena. Finally, I consider ecological modernization theory, a more ideational approach that captures both difference and dynamism, but unfortunately lacks causal impetus. This, the analysis suggests, can at least in part be remedied by looking at institutions and the actors associated with them.[6]

Regulatory Convergence
The notion of "globalization" has taken a strong hold on the political and economic imagination of both the media and the academy in recent years. Debates over regulatory convergence—what it entails, the extent to which it is occurring, and the direction it is taking—have become particularly heated in the twin contexts of European integration and economic globalization. These debates have focused not only on whether convergence is possible, but also whether it is desirable (Vogel 1995). Harmonization (convergence) theory has its roots in international economic theory. In this form, it argues that an increasingly open global economy—one that allows the free movement of factors of production and of goods—will lead to convergence of national differences in prices, rates of inflation, interest rates, and other key economic variables. More "political" approaches examine convergence in governance systems, "the tendency of societies to grow more alike, to develop similarities in structures, processes and performances" (Kerr 1983:3, cited in Unger and van Waarden, 1995:3). In its strongest form, it implies the convergence of countries' policies and policy practices toward identity (Jacobs 1994:32; Hollingsworth, Schmitter, and Streeck 1994). In addition to the pressures imposed by the global economy, Weale et al. (1996:257–

258) identify other pressures for convergence, including issue area characteristics and bureaucratic constraints.

Concerns about convergence on the lowest common denominator are addressed in Vogel (1995). He argues that whether the "California Effect" (convergence toward the highest standards) or the "Delaware Effect" (a "race to the bottom") dominates depends on the "critical role of powerful and wealthy 'green' political jurisdictions in promoting a regulatory 'race to the top' among their trading partners"—a role that has been played by Germany in the EU (Vogel 1995:6). Also important is the degree of economic integration that exists within the trade area. Hence "trade liberalization is most likely to *strengthen* consumer and environmental protection when a group of nations has agreed to reduce the role of regulations as trade barriers and the most powerful among them has influential domestic constituencies that support stronger regulation" (ibid.:8).

Especially since the signing of the Maastricht Treaty on West European Union, the EU has sought harmonization of member states' environmental policies as a way of removing obstacles to free trade, improving transparency, and minimizing environmental damage.[7] The scope of supranational involvement in the domestic affairs of its members has broadened considerably. During the early-1990s, the EU actively pursued the formation of a common European environmental policy, advocating convergence well beyond minimalist notions of convergence of goals and indicators. Indeed, the EU vision appears to be one of harmonization of the very procedural aspects of environmental regulation—structures and styles—studied here (Buller, Lowe, and Flynn 1993). One of the keystones of policy harmonization is the Integrated Pollution Prevention and Control (IPPC) Directive of 1996.[8] This directive is said by many to draw strongly on Britain's IPC legislation (Skea and Smith 1998:265) and is "concerned primarily with environmental procedures (how and under what conditions authorization for industrial sites might be granted) rather than with environmental objectives" (ibid.:278). In its ultimate form it came close to the British model, according to Skea and Smith, in part because British legislation was already in operation and in part because the British model fell close to the median point between the two EU extremes—Germany's strict standards and controls and Spain's desire only to set environmental standards.

"Bottom-Up" and Conflict-Based Approaches
At the same time, several weak points have emerged in the European environmental project. States have made efforts to protect national customs and practices in the face of harmonizing forces, and both the European Environment Agency and DGXI, the environmental directorate, have been relatively marginalized within the EU's governing structures.[9] There are some other strong arguments against the utility of the harmonization debate as a framework for understanding environmental policy change in the EU member states. Among the most cogent is the question of what the EU countries are supposed to be converging to— what is the European regulatory style? Also, as I have shown in previous chapters, regulatory harmonization—at least in its strong form—is not happening among the EU member states. Some states, such as France, have attempted little programmatic restructuring of their environmental policies per se, while Britain's policy élites have been more concerned with restructuring environmental regulation toward a more streamlined, and potentially more easily monitored and managed, system. Germany has been more technocratic in its approach, basing its recycling and reuse policies on the ability of its industries to adapt to new policies through technological innovation within the existing regulatory context. Existing differences between styles and practices have also remained remarkably persistent in the face of pressures for change. In turn, this raises issues of agency: if differentiated responses are to be understood, then the actors (not simply the forces) for change must be specified and their impact laid out. This is where some of the more actor-based theories of EU environmental integration come in.[10]

Several scholars and analysts have taken a "bottom-up" rather than "top-down" perspective on European environmental integration (Lowe and Ward 1998; Liefferink and Andersen 1998; Weale 1995). These works emphasise the interactive nature of domestic-international policy processes, looking, for example, at the way some member states have influenced EU environmental policy directives (Skea and Smith 1998; Jordan 1998a) and at the unevenness of policy implementation. With respect to the latter, recent papers discuss the utility of the logic of "appropriateness," examining how and whether member states adopt EU Directives and if they change or adapt their own systems in accordance with pressure from Brussels (Knill and Lenschow 1998; Héritier 1998). This in turn fits with Skea and Smith's interpretation of IPPC.

Héritier argues—counter to Vogel's contention that states are converging upward toward the German model of environmental regulation—that approaches at the EU level to environmental policy have differed according to issue area. Different areas of European environmental regulation are characterized by measures modeled after the regulatory style of different member states. Thus "in the field of clean air policy, some Directives are shaped according to the German tradition geared towards technology-based emissions control while others are patterned after the British model of regulating ambient air quality" (Héritier 1996:149). Which approach dominates depends first on which state makes the "first move," and subsequently on patterns of problem-solving, "negative coordination, bargaining and compensation" (which determine whether the "first mover" advantage will be translated into policy outcomes) (ibid.:150). Therefore there is no dominant national tradition in European environmental regulation. Instead, what has emerged is a complex and fairly haphazard patchwork of different methods of regulation reflecting distinctive national regulatory styles.

Another set of approaches falls under a "conflict/stalemate" rubric. Although institutional changes introduced under the 1990 Treaty on European Union (the Maastricht Treaty) severely restrict the individual veto power of member states, conflict remains a hallmark of member state interactions in the broad sphere of environmental regulation and has at times threatened the process of European integration itself. For example, in 1995 Shell Oil attempted, with British government approval, to dispose of one of its smaller, redundant oil rigs—the Brent Spar—at sea (Weale 1995). The subsequent vocal opposition from other European countries (including the bombing of a gas station in Germany) and their governments eventually forced the British government and Shell to back down, and the wastes were towed to shore. Bad feelings lingered on to reerupt some six months later with the BSE (Mad Cow Disease) scandal. Following the announcement that links had been established between the bovine and human forms of the disease, British beef exports were banned in Europe, to great outcry from the British beef industry. The British government responded by announcing its intention to block every piece of European legislation until the ban was lifted: this threat eventually led the Council to reconsider the ban.

Environmental issues have proven not to be the technocratic problems that some perhaps hoped. Instead they are highly politicized in the European context, and attempts at further policy harmonization are under threat. The "tension between the international dimension of pollution control and the national basis of environmental regulation" has not yet been resolved (Weale 1995:20). Governments still have both the ability and the desire to resist policy changes from above, albeit sometimes by unorthodox means, and the failure on the part of the EU to resolve conflicts before they erupt into public confrontation underscores weaknesses in the European integration process itself.

In the late-1990s, Brussels took a more cautious approach in handling the more powerful and contentious member states (Haigh and Lanigan 1995:34) as other concerns began to dominate EU politics. Information made available on the implementation of EC environmental laws shows that "suspected infringements of EC rules were higher in the environmental field than in any other area of law except the internal market," and that there is evidence that the Commission "may be slackening the pressure on the member states" to conform with its directives (ENDS Report 261, October 1996:39).[11] Environmental policy was pushed lower in the agenda in the mid- to late-1990s as the EU began concerning itself more with lateral expansion—to include East-Central Europe—and the broader economic integration process. As Andrew Jordan argues, "environmental policy in the EU suffers from two significant 'gaps': an implementation gap—the failure of member states to put its mandates into action—and an integration gap—the failure to incorporate environmental decisions into decision-making at all levels" (Jordan 1998b:39). Neither has been addressed in an institutionalized context.

The theoretical framework developed in this book has much in common with both the "bottom-up" and the resistance scenarios. Britain and France have been sluggish innovators and even in more statist

Germany, reforms have been slow. National—not supranational—actors, policies, and practices have tended to dominate in every case examined here. However, the story is not so simple. Policy innovations have been introduced, and they are quite sweeping. There are also striking similarities between the reforms underway in Britain, Germany, and other EU members.[12] Not only are these reforms roughly coincident in timing, they also reflect a more programmatic, plan-based approach on the part of national governments and demonstrate a shift in policymakers' perceptions about environmental issues—an ideational convergence of sorts.

Ecological Modernization Framework
Knill and Lenschow (1998) introduce the distinction between different modes of impact of EU integration, contrasting command-based impacts —for example, EU Directives—with diffusion-based impacts, such as the transmission of policy ideas among members via the common forums provided by the organization. Following the latter path, another framework captures both the differences and the dynamics of policy change among similar, tightly integrated states, and has been applied across many OECD countries, not just EU member states.

"Ecological modernization" (EM) is a theoretical paradigm that has in recent years become known in the fields of policy analysis and environmental sociology, albeit more in Europe than in the United States. Interpretations of the concept differ vastly, but Weale (1992a) adopts a policy-oriented interpretation.[13] He argues that the first wave of environmental protection—the "old" politics of pollution—shows distinct characteristics across countries. First, most policies separated and addressed pollution control problems by medium: air, water, and land. Second, policy measures tended to be aimed at the point of their greatest effect—in many cases (solid and hazardous waste generation being a case in point) the local level—rather than at the point of generation: a reactive rather than anticipatory approach. Third, policy instruments were based on traditional command and control techniques—for example, setting and enforcing uniform standards across industries.

The start of the second wave of environmental protection—the "new" politics of pollution—coincided with the release in 1987 of the Brundtland Commission Report on Sustainable Development. The most evident changes in regulatory philosophy make up a new series of linkages that together are usually considered to comprise ecological modernization. Its central claim, as developed in policy programs, is "that environmental protection should not be regarded as a burden on the economy but as a precondition for future sustainable growth" (Weale 1993:207). Thus it views recent changes in national regulatory styles and structures as, at least in part, a shift in perceptions of both environmental problems and the most effective means for achieving environmental protection. However, as Weale and Christoff point out, this basic premise has generated a range of corollaries.[14] These can be summarized as follows:

- Economic growth/development and environmental protection are seen as compatible rather than competing goals.
- The environment is more than the sum of its parts (reduction of emissions into one medium can result merely in a shift of the pollution burden to another); hence, effective pollution control depends on an integrated rather than medium-based approach.
- Effective approaches to pollution control take into account effects beyond the local and even the national levels—thus including degradation of the global commons. Most countries have joined international environmental treaty arrangements and compensate for transboundary pollution in their policy measures.
- Anticipation is better than cure. "End of pipe" technologies should be replaced by an emphasis on resource efficiency and recycling and waste control/minimization further upstream in production processes; hence, an emphasis on "green" technology.
- Market- or incentive-based regulatory mechanisms (green taxes, tradable permits, and so on) are superior to traditional command- and control-based regulatory mechanisms. (This so far is an issue more of lip service rather than direct action.)

The main advantage of the ecological modernization thesis over regulatory convergence arguments is that it captures more accurately both ongoing dynamics in the sphere of national environmental policy in recent years and the diversity of national practices and styles. While the ideas driving these changes are often transmitted from the international level—for example, through EU policy forums and consultative processes—the eventual form they have taken on in each country has depended on configurations of domestic-level factors.[15] Therefore an approach that has identified relevant actors and institutions is well placed to apply this framework, and indeed to remedy its major weakness as a theory: its underspecification of the modes and mechanisms whereby certain ideas are transmitted and/or selected by relevant actors.

There are many different groups of stakeholders in environmental policy reform, some of whose interests are more clearly defined or affected than others'. Alignments between these groups, as well as existing consultative and administrative structures, have played an important part in mediating the transmission of new ideas about environmental policy. For example, the British waste disposal industry has consistently supported a more centralized system of environmental management. Institutional factors—which determine, for instance, who has access to the policy process—helped determine the outcome in each case. The more closed British system (hardly surprisingly) remained unresponsive to demands for more open processes of consultation. On the other hand, the Green Movement in Germany—both through die Grünen and through the more established parties and other environmental NGOs—were able to push through a more extensive series of environmental measures that many producers regard as overly restrictive. Domestic institutional factors have also played a key role in slowing down the pace of change. In most of these cases—Britain, in particular—the process of ecological modernization is proving a slow road indeed (Jordan 1993). Years can elapse between the announcement of policy plans and their implementation, and ambitious policies are more often than not severely diluted en route. In other words, the gap between rhetoric and reality remains (Aarhus 1995; Wintle and Reeve 1994).

Despite its breadth, the ecological modernization framework is a more powerful account of changes underway at the level of the EU member states than theories based on regulatory convergence. Viewed through the lens of European environmental policy as a separate issue area, it can be argued that EM provides a more productive guiding strategy for EU policymakers to adopt. Not least, it eschews the discourse of harmonization—which many states still find threatening to their national sovereignty—yet still promises a high degree of environmental protection, and it commands much support from industry and other societal groups. Admittedly, this is reasonably unlikely to happen as long as policymakers are still caught up in the broader rhetoric of European integration and world trade liberalization, which continues to view regulatory differences as a barrier to the free movement of goods and factors of production. However, it is clear that the complex dynamics and interaction between "domestic" and "international" political change will remain an object of study for years to come.

Future of the Waste Trade: Observations and Prescriptions

The international trade in hazardous wastes is a problem that has received much attention in the last ten years: from international organizations such as the UNEP, the OECD, and the EU, the international media, NGOs such as Greenpeace, national governments, and scholars. As I argued in chapter 1, much of this attention has focused on the North-South aspects of the trade and on the trade as a transboundary environmental issue. The international regime governing the waste trade is now moving toward a ban on the transfrontier movements of all wastes from North to South, even those destined for recycling, and the EU is attempting to implement the principle of self-sufficiency in hazardous waste management on the part of its member states. In the following sections I examine the challenges facing the international waste trade regime and the emerging changes, challenges, and opportunities in the field of waste management. In the final sections I identify some of the main flaws in this regime and make some policy recommendations at the international and domestic levels for managing this dynamic issue area.

Reframing the Issue: The State of the Waste Trade
The underlying view of the waste trade presented here is that it is best seen as the most visible symptom of a crisis facing most developed countries: the adequate disposal of the increasing quantities of hazardous wastes they generate. The trade itself emerged as a result of a combination of factors: the expansion of international trade following the lowering of barriers to trade among nations, the vast expansion in the generation of waste products, and the increasingly stringent and

costly regulations placed on waste disposal in many countries. However, the distinctive patterns the trade took on, in terms of disposal routes, depend in turn on national regulatory practices. The heart of the problem lies not in the actual transfrontier movement of wastes but in the need to ensure adequate disposal and regulatory infrastructures—not only in developed countries, but also in countries with emerging or transitional economies.

The movement of wastes among OECD countries cannot be explained by recourse to the pollution haven hypothesis. The strong version of the hypothesis states that firms are likely to relocate in order to take advantage of less strict environmental regulations than those in their home country. Weaker versions of the hypothesis make the claim that goods rating low on the environmental scale—such as wastes and pesticides—are more likely to be exported to such countries.[16] One of the problems with this hypothesis is that it assumes a fairly simple dividing line can be drawn between "weak" and "strong" systems of environmental regulation. Although this is probably the case when comparing industrialized countries with those of the so-called Third World, it is less useful as a benchmark for comparing the industrialized countries with each other. As the analysis I have presented shows, it is not differences in regulatory *effectiveness* that matter here, but rather *procedural* differences in terms of regulatory styles and structures—differences that are much harder to quantify on a simplistic basis. The waste trade first emerged in the absence of any recognition of it being a problem, and it continued that way for many years—until the late-1980s and early-1990s—by which time patterns of trade were well in place. Hence the waste trade can be viewed as an unintended spinoff of regulatory practices that reveals much about the existing regulatory climates of importing and exporting countries. More to the point, the international regime governing the trade—not only the Basel Convention, but also EU and OECD rules and directives—is itself based on the pollution haven hypothesis: that wastes will move from countries with strong regulatory systems to those with weaker systems. While this is certainly the case with respect to illegal waste dumping, it does not apply to the bulk of the waste trade; hence, basing a regime on these principles alone may ultimately prove extremely counterproductive.

Waste Management on a Global Scale: Changes, Challenges, and Opportunities

Trends in Hazardous Waste Generation and Disposal Techniques As tables 2.2a and 2.2b show, the total amount of hazardous wastes generated among OECD countries has increased quite substantially in the years since 1989, despite verbal commitments made by most states to implement waste minimization policies. At the same time, new types of wastes, often posing higher degrees of hazard and/or special handling requirements, are being produced (Wynne 1987). Recent reports show, too, that hazardous waste generation and disposal in newly industrializing, emerging, and transitional economies is increasing to the point where it is becoming a problem due to the lack of adequate facilities and management practices. In 1998, barrels of wastes containing highly toxic substances were discovered in Cambodia. They were found to have originated in Taiwan, which was forced to repatriate the waste after the U.S. refused to accept it.[17]

Actual trends vary internationally. The Russian Federation, for example, has experienced a decrease in amounts of hazardous wastes reported as produced, but is facing a disposal crisis, especially in dealing with the billions of tonnes of wastes still stockpiled. Middle Eastern countries together produce over 1 million tonnes of hazardous wastes per year; outside of Saudi Arabia, however, disposal facilities are minimal, if they exist at all. The picture is similar in Southeast Asia. Malaysia, for example, has doubled hazardous waste volumes between 1984 and 1994. The situation in English-speaking Africa is possibly worse. The International Maritime Organization estimates that these countries generate about 2.23 million tonnes of hazardous wastes annually, over half of which is produced by South Africa. The report notes that "the wastes are mainly (with the partial exception of South Africa) discharged to sewers, sent to municipal landfill or dumped on open land. Of the countries surveyed, South Africa, Namibia and Mauritius appear to be the only countries with commercial hazardous waste disposal facilities." Yet in the figures quoted, Nigeria, Ethiopia, Kenya, and Zimbabwe rank directly behind South Africa in terms of waste generation.[18] In many countries, waste disposal issues are a far more pressing concern than the much more abstract issues of global warming or ozone depletion: "Most of Africa, the Indian subcontinent and Latin America have no waste-water treatment facilities; raw human and industrial sewage is discharged directly into the same bodies of water used for drinking.... In China, an estimated 25 billion tons of unfiltered industrial pollutants went directly into the waterways in 1991, which means there was more toxic pollution in that one country than in the whole of the Western world" (Easterbrook 1994).

Another issue of concern is the continuing problem of developing, siting, and operating new disposal facilities. There is little sign that the NIMBY phenomenon in developed countries is diminishing, and the EU is becoming increasingly strict on landfill requirements. An EU Directive requires operators of existing landfills to report on bringing facilities up to standards by 2002, and by 2004 codisposal will be banned (ENDS Report 280, May 1998:21). Ultimately, landfill will be phased out altogether.[19] More optimistically, there are signs of technological innovation on the part of the waste disposal industry in developing new and safer techniques for disposing of hazardous wastes. These techniques—several of which I outlined in the context of the Australian case—include employment of natural substances to break down toxic elements within a controlled environment and the development of portable disposal facilities that reduce the risks imposed by transporting wastes. In addition, some of the large multinationals in the industry are developing highly sophisticated waste-to-energy recovery facilities. These new techniques are still subject to some problems: they are costly, especially in terms of start-up capital requirements, and in many cases they are still in testing stages.

Trends in the Waste Disposal Industry The waste disposal industry continues to change in scope and structure. In fact, the recent evolution of the industry in response to changing national and international regulations generates some observations pertinent in this context concerning the interactive effect between industry structure, activities, and goals and changing national and international regulatory requirements and policies. The industry has transformed from a conglomeration of many small firms, operating primarily within national borders and often at an extremely local level, to one that is dominated by five or six large and highly diversified multinational corporations. These corporations and their representative trade associations are trying hard, with a moderate degree of success, to redefine themselves—to replace the dominant public image of environmental villains with that of important and socially responsible members of the environmental services industry.[20] To that end (and admittedly with a high degree of private self-interest), they are actively lobbying at a variety of governmental levels against the activities of smaller firms and are engaged in extensive public relations exercises to demonstrate their awareness of their environmental responsibilities. These changes coincide with new assertiveness in business regulation. Cross-sectoral and multinational voluntary codes—including EMAS and ISO 14001, internationally recognized agreements—have flourished in recent years, especially in Europe and in firms dealing with European counterparts.[21]

The industry evolved at first in response to changing national regulations that imposed stricter requirements on waste disposal contractors. In some countries, such as Britain and the U.S., the era of privatization and industry deregulation of the 1980s enabled and encouraged increased private sector activity in this field. These trends in part explain the emergence of the international hazardous waste trade. Concern over the expanding waste trade in turn prompted international regulatory authorities to take action, leading to the negotiation and implementation of the Basel Convention and a set of EU Directives and policy programs aimed at halting the waste trade. At the same time, most governments of industrialized countries pledged to halt exports of wastes to countries not equipped to handle them appropriately and, in a few cases, to ban waste imports. These developments mark the regulatory baseline set in the early-1990s: increasingly stringent disposal requirements at the national level, the emergence of international regulatory authorities onto the scene, and policies in place to ban the waste trade.

The waste industry was quick to respond to these changes, leaving the regulators to catch up. The industry has attempted, unsuccessfully so far, to influence the course of the negotiations over closing the "recycling loophole" at subsequent meetings of the parties to the Basel

Convention. Their influence is on the rise, however, in more recent Technical Working Group meetings to sort out the details of disposal requirements (Clapp 1999). Many big firms, such as Waste Management International and the Danish firm Kommunekemi, are now following relocation strategies, in part to overcome bans on wastes crossing national frontiers and in part in response to demands by many countries for better waste disposal infrastructures (Puckett 1994; Clapp 1997). International bodies have yet to respond to this development.

The waste disposal industry has been more successful in having its demands met by the EU, and there has been a definite change in firms' target level for lobbying activities. It could be argued that they are now setting the pace that government and official actors have yet to reach. The big firms, such as Générale des Eaux and WMI, have been able to take advantage of an expanded opportunity structure for companies offering the sorts of services they do. These include high-level recycling, incineration, and the capacity to collect wastes over a large territorial area. They are among the few actors able to meet the increased stringency of EU and national regulations; however, as Brusco, Bertossi, and Cottica (1996) argue, the EU imposed these regulations only in light of the fact that these firms were able to meet the new requirements.

Future of the Waste Trade: Responding to Regulatory Challenges

The two main bodies engaged specifically in regulating the waste trade at the international level are the UNEP, which administers the Basel Convention and subsequent amendments, and the EU. After a long debate over whether wastes should be classified as a normal good, the EU is increasingly trying to restrict their movement both out of the Union and among the member states. Its support alternates between the proximity principle and the self-sufficiency principle. Individual governments are also taking unilateral steps to address these issues, formulating policies that specifically address the waste trade and problems relating to waste management and administrative structures: it is clear that they are not blind to these problems or (most of) their causes. Causes vary extensively from country to country and are in many cases contingent upon preexisting national circumstances. There is no blueprint for effective waste management, but individual countries can learn much from each other by adapting new technologies and regulatory approaches to fit their particular circumstances. Coordination and facilitation at the international level of such activity is needed not only for less developed and emerging economies, but for the most advanced economies as well.

The UNEP and the EU differ significantly in their powers to enforce compliance with their rules and recommendations. The EU oversees a small group of highly industrialized nations that share a high level of common interest, and it has considerable authority to intervene in the affairs of its member states. The UNEP, on the other hand, has the unenviable task of coordinating the interests and demands of all UN member states. Hence the waste trade regimes of these two bodies differ significantly in scope and means, while sharing the same ultimate goal: to stop waste trading, especially the export of wastes to countries ill-equipped to handle them. Hazardous waste policy in the EU is highly integrated with the overall package of policies that cover environmental policy and the movement of goods around the Community, while the UNEP engages in much less issue linkage across international environmental agreements.[22]

In both cases there is an overriding need for a commonly accepted definition and classification of what exactly constitutes a hazardous waste, and there needs to be an international clearinghouse for data on all aspects of hazardous waste management.[23] Recent negotiations toward waste lists—under the auspices of both Basel and the EU—are moving slowly, as recent reports of the Technical Working Group of the Basel Convention suggest, although the OECD's red, amber, and green lists of wastes have provided a model (OECD 1993b). Without reaching agreement on these factors, these regimes will not even achieve the minimal goal of improving issue transparency. These organizations might do well to learn from the institutions in place under NAFTA. While less than perfect, data tracking systems such as Haztracks provide useful data on the process for and extent of waste transfers between the United States and Mexico.[24]

The Basel Convention does not directly apply to the legal waste trade among OECD countries; nonetheless, the findings presented here have important implications for the ultimate effectiveness of that regime.

First, it fails adequately to address the underlying causes of the waste trade—in particular, the waste management crisis in developed countries. Even though waste volumes are increasing across countries and continue to outstrip available disposal capacity, the Basel Convention is one of the few international agreements aimed at regulating polluting substances that does not establish targets for reducing the production of the pollutants in question. Failure to address these issues—especially given the relative lack of monitoring and enforcement capabilities—could lead to one or more of several possible outcomes:

- A continuance and escalation of illegal waste dumping on less developed countries. There have been many recent reports of U.S. wastes being dumped in China and Hong Kong, and Germany recently had to take back a shipment of wastes illegally sent to Lebanon.
- Defection from the Basel Convention by leading players. Countries that have already expressed an unwillingness to comply with the ban on global movements of scrap metals include India, Australia, and the United States.
- Diversion of trade to more developed countries willing to receive these wastes. These could include "second-tier" emerging economies, such as in East Central Europe, Southern Europe—figures show a vast increase in waste importation by Spain in recent years—and East and Southeast Asia. As analysis of Britain and France showed, countries that are highly developed do not necessarily have disposal facilities capable of destroying toxic wastes in the safest manner.
- Relocation or expansion of waste disposal firms abroad.
- Costly and potentially highly dangerous storage of wastes in countries unable to dispose of them in a timely fashion. This has already been seen to be the case in Britain and Australia.

A similar set of results could emerge should the EU adopt and enforce the principle of self-sufficiency in waste disposal among its member states. France appears to have opted out of this process. Some states, such as Germany, are objecting to the inclusion of a ban on the transfrontier movement of wastes destined for recycling or recovery. Other states, such as Britain, are finding it hard to translate the self-sufficiency principle into practice.

These are not easy issues to address. However, some basic recommendations can be made. First, the international community must seriously address questions of waste generation by both advanced industrialized and less developed and emerging economies and begin to set targets for reducing the volumes of hazardous wastes generated for final disposal. Some countries, such as Germany, are beginning to implement effective waste minimization policies. These appear to work best when attention is focused on production processes themselves—especially the reuse and recycling of materials within the manufacturing process—and when new, more efficient technologies are employed. Again, the private sector appears to be ahead here, as data on the growth of the environmental services sector shows. Such measures also appear to command a high degree of support from broader societal interests.

Second, effective and lasting waste minimization practices are a very long-term goal that will require a high degree of reorganization and restructuring of existing industrial practices and attitudes. There are, however, some interim solutions emerging. As outlined already, large strides have been made in recent years toward the development of new waste treatment technologies, including more environmentally sound techniques, the development of portable treatment plants, and the development of technologically advanced waste-to-energy recycling and recovery plants. However, the political will is lacking to put many of these innovations into practice or to provide the sorts of subsidization needed, especially at first, to enable firms that wish to use them to cover start-up costs. This is the sort of research and development activity that could be undertaken by bodies such as the EU and the UNEP, who are able to draw on a wide range of expertise from their member states.

Third, as this work demonstrates, broader regulatory and institutional structures matter just as much as specific rules and regulations regarding hazardous wastes in seeking to control the trade. Several countries are beginning to take steps to address existing flaws—administrative reorganization in Britain, for example. This observation applies equally well to the development of waste management infrastructures in developing, transitional, and emerging economies as it does to the industrialized democracies of the world. This is becoming a high priority for many international organizations and funding institutions such as the World Bank, the World Health Organization, and USAID.[25] Existing studies highlight many of the obvious problems, from differing industrial structures to the lack of many important needs such as reliable information on the problem, financial resources, technical know-how, and adequate transportation.[26] A common conclusion is that "there is no single control system for hazardous waste that will work perfectly in all countries. The legal, political and cultural system in each country demands a unique national solution" (Forester and Skinner 1987:16), as indeed is the case in the OECD.

Some specific recommendations for the development of regulatory infrastructures in these countries emerge from this work nonetheless. First, administrative structure is extremely important: in order to control issues—such as hazardous waste management—that have complex local, national, and international ramifications, clear chains of communication between government agencies and administrative units are necessary. Second, it is vital to include as wide a range of interests as possible in the environmental policy process, especially in the immediate policy community. Finally, an implementation process that covers, or at least monitors, the "cradle-to-grave" life cycle of wastes is very important. Perhaps then, at the very least, these countries could learn from some of the mistakes made by their more industrialized counterparts.

Conclusion

Overall this work predicts that the waste trade—in particular the trade among the OECD countries—will not vanish in the near future, despite the best efforts of international regulatory authorities, and that a blanket ban on waste trading would be highly unlikely to be effective. There are several reasons why this is likely to be the case. International relations theory predicts that cooperation is unlikely to work when states disagree over the basic principles of environmental agreements and international authorities lack effective enforcement capabilities.

Most basically, the national regulatory differences driving specific patterns of waste trading remain, and are being eroded slowly, if at all. Furthermore, little concerted and conscientious effort is being made to address the waste trade in the context of the hazardous waste management crisis facing most, if not all, industrialized countries. Paradoxically, it in fact seems that the only actors seriously taking on board notions of environmental responsibility are the large waste disposal firms themselves, who favor (with an eye on potential profits, of course) the establishment of an international network of high-tech treatment plants and transportation routes.

A theme that has come up time and again in the empirical analyses is that of the NIMBY phenomenon: communities generally oppose the imposition of waste disposal facilities in their immediate localities, and historical experience with practices of uncontrolled landfilling has shown these fears to be justified. Many argue that the bonds of trust between local communities and regulatory agencies have been broken (Wynne 1987; McDonell 1991; Munton 1996a). The literature on waste exportation shows that the NIMBY phenomenon has been globalized and indeed has come to be practically an official policy position for several governments. There is, therefore, a distinct need for most societies—at the individual, local, and government levels—to accept responsibility for waste generation practices and for the adequate disposal of these wastes. This implies much better communication of potential risks to achieve a shared understanding and will require high levels of government transparency, greater involvement of community groups in decision-making processes, public education programs, and official commitment to searching out and employing the best available disposal technologies and regulatory practices. New and creative work on these subjects bridging the policy-academe gap, such as Rabe, Gunderson, and Harbage 1996a, and Munton 1996, needs to be built upon.

Finally, the effect of different national responses to environmental degradation and to international commitments to change behavior is a subject that is understudied in the field of environmental politics, both comparative and international. As at least a first corrective step, I have demonstrated that institutional constraints and opportunities for actors and stakeholder groups affect the interests and behaviors of those institutions, actors, and stakeholders. In addition, I have established the importance of these same forces on political outcomes at the local, national, and international level. Further research is required to apply this model of comparative regulation to other countries and public policy issue areas—both environmental and otherwise—to capture the full range of dynamism and difference encapsulated in these national regulatory systems.

Notes

Chapter 1

1. See Montgomery 1995:4.

2. For broad discussions of institutional approaches to comparative politics, see Steinmo, Thelen, and Longstreth 1992, Cammack 1992, Ikenberry, Lake, and Mastanduno 1988, Elman 1995, and Hall 1997.

3. Thompson's article opens with the observation that "few rewards come the way of those who take rubbish seriously" (Thompson 1994:199), based on Swift's late-seventeenth-century remark that "social status and intellectual respect are accorded to the person who treats rubbish with the 'correct' circumspection" (Swift 1696).

4. For discussions, see Strohm 1993 and Montgomery 1990 and 1995.

5. For overviews of some of the debates surrounding institutionalist theory, see Cammack 1992, Steinmo, Thelen, and Longstreth 1992, and Elman 1995.

6. According to the World Resources Institute, over 170 international environmental treaties have been adopted, more than two-thirds of them since the 1972 UN Conference (World Resources Institute 1994).

7. International environmental regimes are best defined as "sets of international regulations and organizations which were intentionally established by preexisting actors (states) through explicit, legally binding or politically binding international agreements, in order to regulate anthropogenic sources of negative externalities affecting the international environment" (Bernauer 1995: 352).

8. See Zürn 1998 for a similar argument.

9. See VanDeveer 1997 and Weinthal 1998, for instance.

10. For examples of works that problematize the relationship between IR theory and the environment, see for example Vogler and Imber 1996 and Redclift and Benton 1994. For other works critical of state centricity in global environmental politics, see Lipschutz and Conca 1993, Lipschutz and Mayer 1996, Wapner 1996, and Low and Gleason 1998.

11. See on this point Young 1990, Hurrell and Kingsbury 1992, Porter and Brown 1995, Hahn and Richards 1989, and Stevis, Assetto, and Mumme 1989. According to Young, "The central concern of such studies is to identify the determinants of success or failure in efforts to form international regimes; some attention has been devoted as well to explaining the timing of successes in regime formation and to accounting for the content of principle provisions of the regimes that form" (Young 1990:340). Two of the most interesting approaches are Peter Haas's theory of epistemic communities (Haas 1990a and 1990b) and Young's model of institutional bargaining (Young 1989), which emphasizes the role played by the existence of uncertainty. Also important is the role of entrepreneurial leadership (Benedick 1987).

12. These are: (1) Effectiveness as Problem Solving, i.e., "whether regimes operate to solve the problems that motivate parties to create them in the first place"; (2) Effectiveness as Goal Attainment, "a measure of the extent to which a regime's (stated or unstated) goals are attained over time"; (3) Behavioral Effectiveness, "whether the operation of a regime causes one or more of its members ... to alter their behavior either by doing what they would not otherwise have done or by terminating or redirecting prior patterns of behavior"; (4) Process Effectiveness, "the extent to which the provisions of an international regime are implemented in the domestic legal and political systems of the member states as well as the extent to which those subject to a regime's prescriptions actually comply with their requirements"; (5) Constitutive Effectiveness, "in the sense that [a regime's] formation gives rise to a social practice involving the expenditure of time, energy and resources on the part of its members"; and (6) Evaluative Effectiveness, "whether the regime produces results that are efficient, equitable, sustainable or robust," based on specific performance criteria such as cost effectiveness (Young 1994:143–149).

13. See Mitchell 1994:332 for an elaboration of the distinction between compliance and effectiveness. Yet it is possible to think of cases where the set of institutional rules that best ensure state compliance are not the optimal ones for ensuring effectiveness.

14. See for example Susskind, Siskind, and Breslin 1990, a set of analyses of the negotiations surrounding a number of environmental issue areas. Although such individual case studies often take an atheoretical approach to the issues, they are most useful in providing rich narrative and in examining lessons that can be applied from one issue area to another (Downie 1994; Morrisette 1991).

15. The Convention-Protocol method of environmental negotiation is the model followed by the UNEP in negotiating international agreements. First, countries meet to draft a framework convention—usually a general statement of principles. Subsequent meetings of the signatories lead to the drafting of protocols, which are a more explicit declaration of goals to be achieved and policies followed, and involve a definite commitment on the part of states (Susskind 1994:30).

16. On international nonstate actors, see in particular Litfin 1993, Kamieniecki 1993, Princen and Finger 1994, Wapner 1995, and Fox and Brown 1998.

17. See Mitchell and Bernauer 1998 on problems of qualitative analysis and case study design in international environmental politics.

18. Indeed, as some argue, some agreements are designed to allow for minimal behavioral change on the part of states, making it easy for them to "comply" but having next to no effect on the environment.

19. See Sprinz and Vaahtoranta 1994, Paterson 1996, Schreurs and Economy 1997, and Porter and Brown 1995.

20. For overviews of the field of comparative environmental policy, see Vogel and Kun 1987 and Buller, Lowe, and Flynn 1993.

21. There are, however, several studies that examine environmental regulation in less developed countries. These include Bromm 1990, Leonard and Morrell 1981, Leonard 1985, Handl 1988, Maluwa 1989, Penna 1993, and Desai 1998.

22. For examples, see Blowers, Lowry, and Solomon 1991 on nuclear waste disposal and Brickman, Jasanoff, and Ilgen 1985 on regulating dangerous chemicals. On public policy comparisons, see Heidenheimer, Heclo, and Adams 1990.

23. See for example Vogel 1986, Sbragia 1991, and Moe and Caldwell 1994.

24. Although these are for the most part set in the U.S. context (Buller, Lowe, and Flynn 1993).

25. See Marr 1997 and Barrett and Therivel 1991 on comparative assessment, and Lotspeich 1998 and Aarhus 1995 on comparative use of market-based instruments.

26. For alternative formulations, see Vogel and Kun 1987, Knoepfel et al. 1987, and Lehman 1992.

27. See Inglehart 1982, Adeola 1998, Dunlap and Mertig 1997, and Brechin and Kempton 1994.

28. See Scruggs 1999, Dalal-Clayton 1996, and Jahn 1998.

29. Studies of policy styles can be traced back at least to Richardson 1982, an early and influential work in this field. For an overview of these approaches, see van Waarden 1995.

30. Again, mirroring the broader institutionalist literature, which tends to differentiate between rational choice and historical approaches to the evolution of institutions. See Steinmo, Thelen, and Longstreth 1992.

31. With the exception of Boehmer-Christiansen and Skea 1991, which examines differences in British and German responses to transboundary air pollution.

32. See for example Vogel 1995, Héritier et al. 1994, Lowe and Ward 1998, Caporaso and Jupille 1998, Majone 1996, and Weale 1995.

33. See Zaelke, Orbuch, and Housman 1993, Low 1992, OECD 1994a, Jaffe, Peterson, and Portney 1995, and Thompson and Strohm 1996.

34. Copeland identifies three types of transboundary externalities in the relevant literature. The first involves situations where pollution generated in one

country spills over into another, the second where pollution control policies and tariffs can affect the location of the pollution generating industry—the pollution haven hypothesis—and third, situations where the pollution effects are confined to the recipient country but the generator stays in the country of origin (Copeland 1991:143–4).

35. For a discussion of definitions of wastes and related issues, see Gourlay 1992, chapter 1, and Thompson 1994.

36. For one of the most comprehensive discussions of the politics of risk in hazardous waste debates, see Wynne 1987. For more general discussions of the lay/expert risk assessment debate, see Chapter 2 of this book, Armour 1993, and Portney 1988.

37. See Jasanoff 1986 and 1991 and Brickman, Jasanoff, and Ilgen 1985.

38. Although, see Auer 1996.

39. The relationship between democracy and the environment has been the subject of much work in the field of environmental political theory. See for example Dobson 1995, Eckersley 1992, and Dryzek 1987. For overviews of the New Social Movement literature and environmentalism, see Richardson and Rootes 1995, Taylor 1995, and Dalton and Kuechler 1990.

Chapter 2

1. For comparative studies of national hazardous waste management practices, see Forester and Skinner 1987, Wynne 1987, Munton 1996a, and a special issue of the *International Journal of Environment and Pollution*, 7:2 (1997), ed. H. W. Gottinger.

2. See Gourlay 1992, chapter 1, for a discussion of these issues.

3. This point is made persuasively by Easterbrook (1994), who points out that in many less developed countries, sewage and industrial pollutants are discharged directly into water used for drinking and household activities—the direct cause of many problems for human health.

4. See Dowling and Linnerooth 1987. Throughout the literature, the terms "toxic" and "hazardous" are often used interchangeably. Correctly, "toxic" wastes are those that pose a threat to human health specifically, while the term "hazardous" covers a broader range of substances, including not only those that threaten health but also those that pose a risk to the environment (World Resources Institute 1987:201).

5. This definition mirrors that used by the United States Environmental Protection Agency, which defines as hazardous any waste that meets any of the four criteria of ignitability, corrosivity, reactivity, or toxicity (Maltezou 1989:263).

6. According to Asante-Duah, Saccomano, and Shortreed (1992:1685), high-risk wastes are those "known to contain significant concentrations of highly

toxic, mobile, persistent and bioaccumulative constituents. Examples are chlorinated solvent wastes from metal degreasing, cyanide wastes, dioxin-based wastes, and PCB wastes. Intermediate-risk wastes include metal hydroxide sludges for which the toxic metals are in relatively insoluble form with low mobility. Low-risk wastes include primarily high-volume/low-hazard wastes and some putrescible wastes, for which the cutoff between a 'hazardous' and a 'non-hazardous' waste is least clear-cut."

7. On the waste management hierarchy, see for example Forester and Skinner 1987 and Batstone 1989:97. Other sources used for this section include Allen 1992, Bromm 1990, Handl 1988, Maltezou, Biswas, and Sutter 1989, OECD 1985 and 1993b, and Piasecki and Davis 1984 and 1987.

8. After detoxification, many wastes contain reusable components. For example, in Denmark the major disposal plant supplies a good percentage of the energy requirements of the surrounding area.

9. Incineration is a much more common method in more developed countries (Forester and Skinner 1987:64), and there are various technologies available. The rotary kiln is the most common form, but increasingly countries are turning to the use of cement kilns, a more efficient means of disposal (Forester and Skinner 1987:56). Physical and chemical treatments are also more common in countries with more developed control systems.

10. The most insecure form of landfill is the so-called dilute and disperse site, where wastes are buried in an unlined site with the expectation that wastes will be broken down into less harmful components as they seep through the ground (Hildyard 1986:225). This problem is compounded by the practice of codisposal, unique to the United Kingdom, whereby hazardous wastes are mixed with other types of wastes, following the principle that hazardous elements will be neutralized through reaction with elements present in the other wastes (Postel 1987:13; Piasecki and Davis 1987:193).

11. See Blowers, Lowry, and Solomon 1991, Kemp 1992, Herzik and Mushkatel 1993, Munton 1996a, and O'Neill 1998b for some comparative and international overviews of nuclear waste management.

12. Environmental degradation is defined as "the introduction into the environment of substances or emissions that either damage or carry the risk of damaging, human health or well-being, the built environment, or the natural environment" (Weale 1992a:3). Of particular concern are anthropogenic pollutants—those that have their origins in human activity. Environmental management consists of attempts to slow, halt, reverse, or prevent such degradation. Given that many environmental problems fall within the class of market failures (e.g., externalities or collective goods), such management entails deliberate action, usually, although not always, on the part of government actors.

13. An illustrative example: it was predicted that the Freshkills landfill on Staten Island would by 1998 become the highest coastal point on the eastern U.S. seaboard between Maine and the tip of Florida (Gourlay 1992:90).

14. In accordance with the data presented in table 2.2, most sources agree that hazardous waste production has increased significantly over the past decades, although, of course, they disagree over how much. The UNEP reports that "comparison of the hazardous waste figures for some OECD countries in the early 1980s and the late 1980s indicates that hazardous wastes are increasing. The production of hazardous wastes in the USA, for example, increased from about 9 million tonnes per year in 1970 to about 264 million tonnes at present." (Tolba and El Kholy 1992:265). Yakowitz 1993 states that in OECD Europe, which generates roughly 24 million tonnes of hazardous wastes per year, waste generation is increasing roughly 1 to 2 percent per year (138).

15. There have been many studies on the effects on human health of exposure to toxins found in hazardous wastes. Allen 1992 provides an extensive analysis of the effects of exposure to dioxins on human health, and Bullard 1991 and Seager 1993 provide persuasive evidence of the adverse health effects of dumping on communities both in the United States and abroad. See Heller 1994 for a listing of the environmental and health effects of twenty-four of the most commonly traded forms of hazardous wastes.

16. Helen Dolk, Martine Vrijheid, and Ben Armstrong, "Risk of Congenital Anomalies near Hazardous-Waste Landfill Sites in Europe: The EuroHazcon Study," *The Lancet*, August 1998; "Scientists Link Birth Defects to Landfill Sites," *The Guardian*, August 7, 1998:5.

17. For an overview of Superfund, see Rahm 1998.

18. See for example Bullard 1991, United Church 1987 (a landmark report in this field), Newman 1994, and Faber 1998.

19. For extensive discussion of the NIMBY phenomenon and its roots, see Munton 1996.

20. As well as Armour, see also Cothern 1996, Wynne 1987, Renn et al. 1992, Finkel and Golding 1994, Wildavsky and Dake 1990, Hertzman and Ostry 1996, and Linder 1997 on the social, political, and economic constructions of risk.

21. On the evolution of risk assessment debates in the U.S., see Rosenbaum 1997 and Linder 1997.

22. See Beck 1995 and 1996, Carson 1962, and Blowers 1997.

23. See Buell 1998, an article that uses critical theory to pull these themes together.

24. These are, among other things, institutional issues, and the conclusions they lead to about institutional design are reviewed in the final chapter.

25. See Jasanoff 1997, Powell and Leiss 1997, and Levidow et al. 1997.

26. Wynne terms this an "inverse materials-cash relationship" (Wynne 1987: 76–78).

27. One estimate puts the average cost of waste disposal in developed countries at anything between $75 and $1500 per ton, depending on the type of waste and the disposal method used or required (Montgomery 1990:314), and "for the United States, some 1985 data for the state of New Jersey reported disposal cost ranges of $200 to $2600 per ton" (OECD 1993b:11). According to Hilz and Ehrenfeld, the costs of landfill in the United States has increased sixteenfold since the early 1970s, and the cost of incineration increased threefold between 1980 and 1989 (Hilz and Ehrenfeld 1991:33).

28. For a thorough analysis of factors influencing waste export decisions, see Hilz 1992.

29. Although it is highly likely the trade began well before that date. As a matter of interest, there is a cartoon in the Ellis Island Museum, New York, dating from the 1920s, that compares the flood of immigrants from Europe to a flood of toxic wastes from abroad.

30. There is significant disagreement among waste trade analysts, both activists and academics, over the extent of the illegal trade. While Montgomery (1995) disputes what he terms the "iceberg theory" of waste dumps, other organizations, in particular Greenpeace International, vehemently disagree. The focal points of disagreement are the admissibility of much of the evidence and the question of what actually constitutes an illegal shipment of wastes. There is a large gray area in international law, whereby wastes labeled as "recyclable," and hence legal under international conventions, are often claimed to contain many harmful elements (Puckett 1994).

31. For descriptions of the early and colorful history of the waste disposal industry, see B. W. Clapp 1994, Crooks 1993, and Block 1994. See Cooke and Chapple 1996 for analysis of recent trends in the waste disposal industry: chronologically, privatization, globalization, and concentration. For example, Waste Management International, a British-based firm of American parentage (WMX Technologies) operates in nineteen countries worldwide.

32. A combination of all of these factors contributed to the Bhopal Chemical disaster of 1984, in which several thousand people were killed and hundreds of thousands of others injured through the leakage of toxic gas from a pesticide plant in Central India owned by the American Union Carbide Corporation. A further example of this phenomenon is the *maquiladora* problem, whereby U.S. firms have been able to take advantage of lower factor costs in Mexico but are able to export their goods back to the U.S. without tariffs. This has led to what has been described as an environmental wasteland along the border (see Simon 1997). See also Clapp 1998a on the export of hazard to less developed countries from the industrialized world.

33. Another estimate puts the figure of wastes disposed of outside their country of origin at 15 percent of wastes generated internationally—roughly 45 million tons per year (OECD 1985, cited in Hilz and Radka 1990:76). A recent statement from the UNEP estimates that roughly 10 percent of the 400 million

tonnes of wastes produced globally per year cross-national borders (UNEP Information Note 1997/22, released June 1997).

34. The phrase "domestic dependent nation" was used first by Chief Justice John Marshall [*Cherokee Nation v. Georgia*, 30 U.S. (5 Pet.) I (1831)]. According to one analysis, "[t]hough subject to the guardianship protection and superior political power of the federal government, Indian nations did possess some degree of sovereignty": while not equal to foreign nations, they "did constitute legitimate legal and political entities that could manage their own affairs, govern themselves internally, and engage in legal and political relations with the federal government and its subdivisions" (Deloria and Lytle 1983:4).

35. On the role of NGOs such as Greenpeace and the UNEP in initiating international action on the waste trade, see Clapp 1994 and Montgomery 1994. The latter examines sources of international pressure on Guinea-Bissau in its refusal to accept a waste import scheme that on the face of it appeared environmentally sound. The role of Greenpeace International (and its Toxic Trade Project) in combating the illegal waste trade cannot be underestimated. Not only has it extensively documented such cases, but it has also spearheaded attempts to galvanize public opinion against the trade in many parts of the world.

36. According to Montgomery, and indeed to the Greenpeace figures, "between 1970 and 1990, Greenpeace uncovered 103 proposals to ship hazardous waste to developing countries.... Of the 103 proposals, only 16 resulted in a transfer of hazardous waste across an international boundary" (Montgomery 1995:6). Exports to middle-income countries—notably East Central Europe—over the same period were much higher, with waste transportation occurring in 41 out of 98 proposed schemes (ibid.:8).

37. Some of the wastes shipped abroad veer toward the bizarre: for example, Romania was the recipient in 1990 of one to two thousand tonnes of radioactive tobacco, courtesy of an Italian company (Heller 1994:77).

38. For example, in August 1994 the *Daily Mirror*, one of Britain's tabloid newspapers, ran a front page story under the headline "Ship of Doom" about the proposed offloading in Liverpool of a cargo of contaminated waste from Germany. The significance of this is that coverage of the story was not restricted to the *Guardian*, Britain's most liberal daily. Similar stories, as well as extensive television coverage, ran during the Karin B. affair in 1988, when that ship attempted to dock in Britain carrying a cargo of hazardous wastes from Italy— at first with official government approval.

39. Copeland characterizes the waste trade as a special sort of transboundary externality, where the externalities from waste disposal are assumed to be confined to the country where the waste is disposed of, but which can be transported from the country of origin without having to move the waste producing industry (Copeland 1991:144).

40. See, for example, Vallette and Spalding 1990, Third World Network 1989, Clapp 1994, Crooks 1993, Heller 1994, Singh and Lakahan 1992, Strohm 1993, and Park 1998.

41. These plans were finally revisited in 1998; see "New York Tries to Clean Up Ash Heap in the Caribbean," *New York Times*, January 14, 1998:A5.

42. Krasner defines an international regime as "principles, norms, rules and decision making procedures around which actor expectations converge in a given issue area" (Krasner 1983:1). In this case, Krasner's definition is extended to a regime constituted by multiple and often conflicting international agreements.

43. In 1984 the OECD adopted a Decision and Recommendation regarding intra-OECD trade, which stated that "member countries shall control the transfrontier movement of hazardous wastes and, for this purpose shall adequately ensure that the competent authorities of the countries concerned are provided with adequate and timely information concerning such movements" (OECD Council Decision and Recommendation C (83) 180 (Final)). It was amended in 1991, when it was decided that waste movements should be restricted to wastes destined for recovery operations (C (90) 178/Final). In 1992 the OECD adopted a control system, applied only to OECD members, classifying wastes into green, amber, and red lists, depending on their overall environmental risk and their management practices (OECD 1993b; Yakowitz 1993). This classification determines the restrictions imposed on transfrontier movement of the wastes. Only wastes categorized as red are subject to the full force of international regulation under the Basel Convention; "green" wastes are not subject to any controls, while "amber" wastes require notification and implied (as opposed to written) consent. See Rosencranz and Eldridge 1992:321.

44. Negotiations followed the convention-protocol approach generally adopted by the UN in negotiating international environmental agreements. First, a framework convention is drawn up, establishing general aims and principles of the agreement. Once that has been ratified by the contracting parties, more detailed protocols are drawn up, setting out more specific targets and goals for parties involved. See Susskind 1994 for an analysis and critique of this mode of negotiation.

45. In 1998 the U.S. Environmental Protection Agency announced that it would seek to ratify the Basel Convention in its original 1989 form, a move that has been greeted with opposition from environmental and some industry groups. For further discussion, see O'Neill 1999a, and for updates on ratification, see the website of the Basel Action Network (www.ban.org).

46. For an overview of Basel Convention diplomacy, see Krueger 1999.

47. The key opponents of the recycling ban include the International Chamber of Commerce (whose representative sat on the negotiating committee), the International Council on Metals and the Environment, and the Bureau of International Recycling. Their main demand was that the ban be delayed until agreement had been reached over the term "hazardous." Their views were laid out succinctly in an editorial in the *International Herald Tribune*, October 4, 1995 ("Hurting Development and Business," by John C. Bullock, of the ICC). There is also evidence that a few countries—for example, the U.S., Canada, India, and Australia—supported their position. See Clapp 1999 and Alter 1997

for a full analysis of industry motivations and tactics, which include highly active involvement in the Technical Working Groups associated with Basel.

48. For overviews of developments in European environmental policy, see Liefferink, Lowe, and Mol 1993, Lévêque 1996, Lowe and Ward 1998, and Jordan 1998b.

49. There are two forms of EC legislation: regulations, directly enforceable in national courts under the Doctrine of Direct Effect, and directives, which comprise the majority of community legislation. Rather than being enforced from the supranational level, directives have to be individually implemented by the national legislatures of member states. In the event that the national government fails to implement a directive, the matter is referred to the European Court of Justice.

50. For a summary of EC/EU legislation related to hazardous waste management, see Porter 1998:203.

51. For a detailed discussion of the European debate, see Zito 1994 and Jupille 1996. One of the landmark cases in this area was the Walloon Waste Case of 1992, when the ECJ upheld Belgian legislation banning the imports for waste from other countries and other areas of Belgium into Wallonia. In their rulings on the case, the judges upheld the argument that wastes are a "normal good" and hence subject to free trade rules, but argued that, owing to the self-sufficiency principle, they are "attached to their point of production" (Jupille 1996:10–11).

52. However, the self-sufficiency principle applies only to wastes destined for final disposal; wastes destined for recovery purposes are subject to the same trade rules as normal goods. There is also some disagreement over whether OECD "green list" wastes should also be included in regulations governing the waste trade.

53. While the packaging waste controversy does not deal directly with hazardous wastes, the complicated evolution and unexpectedly controversial nature of this debate highlights the problems that differences in policies and practices among the member states have caused for waste management policy in the EU. The main problems were twofold. First, many states viewed the stricter laws imposed by Germany and Denmark as barriers to trade (see earlier discussion of the German Packaging Ordinance). Second, these countries soon found they had much more packaging waste than they could handle; hence, Germany in particular began exporting its waste packaging material to other countries, such as the U.K., for recycling. This had the effect of undercutting the nascent recycling industry in these countries. After a period of protracted negotiations, the EU responded to pressure from the affected parties by eventually (December 1994) approving a recycling directive that "requires member states to recover at least half of their packaging wastes, and recycle a quarter of it ... within five years," on the condition that domestic recycling facilities had adequate capacity for handling the wastes (Vogel 1995:92). This controversy remains unresolved.

54. An additional difference between the waste trade and classic transboundary issues concerns the question of agency: "classic" pollutants are normally transmitted across boundaries by natural mechanisms, but transporting solid wastes across boundaries necessarily involves human agency, in terms of the decision to transport the waste.

55. See also O'Neill 1998a.

56. On the absence of, and attempts to develop, waste management systems in less developed countries, see Bromm 1990, Batstone, Smith, and Wilson 1989, and Wilson and Balkau 1990. To some extent, the UNEP is sponsoring the development of "regional or sub-regional centers for training and technological transfer regarding the management of hazardous wastes and other wastes and the minimization of their generation."

Chapter 3

1. See Elman 1995 for a discussion of the use of domestic-level approaches in international relations theory.

2. This is one of the main claims of the New Institutionalist literature. For example, see Cammack 1992, Krasner 1988, Steinmo, Thelen, and Longstreth 1992, Ikenberry 1988, Hall 1986, and Haggard 1990. Like these works, this book takes the perspective of "historical institutionalism," a more sociological-historical approach to the effects of institutions on behavior, as opposed to the "contractual," or rational, choice variant. Historical institutionalists argue that institutional structures are not the function of individual choices; rather, choices are filtered through these structures. See Cammack 1992:402–403 and Thelen and Steinmo 1992:7–10 for a discussion of the main differences between these approaches.

3. For an overview, see Majone 1996.

4. This study uses both the volume of hazardous wastes a country imports or exports each year, measured by weight, and figures for net imports (imports less exports). This technique controls for transit states such as the Netherlands—which imports significant quantities of hazardous waste each year, only to reexport it for disposal elsewhere—and also provides a measure of a country's ability to absorb hazardous wastes over and above the amount it produces and disposes of domestically. At the same time, some notion of the order of magnitude of waste movements in and out of countries is required to provide an idea of the extent of a country's involvement in the trade. Hence data presented in the empirical section includes imports and exports of wastes by country, as well as the figure for net imports.

5. See the final section of this chapter for issues arising regarding available data on intra-OECD waste trading.

6. Forester and Skinner identify the main elements in a national hazardous waste control system as, first, waste definitions, then "responsibilities placed on

the waste generator and registration or licensing of those involved in collection, transport, intermediate storage, treatment, and disposal. Other elements are control over transport; permitting of treatment or disposal facilities; and programmes for dealing with old or abandoned sites" (Forester and Skinner 1987:16). However, they and Wynne (1987) show that while these elements exist in most waste management systems, their substantive content, interpretation, and means of implementation differ significantly across countries.

7. Environmental policy in industrialized countries tended to develop along medium-specific lines, i.e., water, air, and ground pollution. However, many countries have adopted, or are considering adopting, systems of integrated pollution control, defined as "the range of organizational and legislative changes that enable institutions to deal with the connected nature of environmental problems" (Irwin 1990:9). There are two ways in which integration can occur. The first is "internal," whereby the environment as policy problem is treated as a unified whole, and the second is "external," whereby environmental policies are integrated into all sectors of public policy (see Haigh and Irwin 1990, Jordan 1993, and Weale, O'Riordan, and Kramme 1991). As shall become clear in the empirical analysis, some national systems are more integrated than others.

8. These are 1) the public goods nature of environmental problems, 2) the externality effects of pollution, 3) the large technical core of environmental policy, 4) that their effects typically occur over the long term, and 5) that environmental issues cut across the established sectors of public policy. In this analysis he draws on Lowi (1964), who argues that policymaking across different issue areas is driven by the particular characteristics of each area (Weale 1992a, chapter 1).

9. However, this is not the case worldwide: as observers point out, many less developed countries have no national system of environmental regulation, much less regulations pertaining to the disposal of hazardous wastes (Bromm 1990; Maluwa 1989; Leonard and Morell 1981). Alternatively, although a country's constitution may state the importance of protecting the environment, there may have been no laws enacted to implement it, as, for example, in Ghana (Maluwa 1989:661). On the other hand, some OECD countries—notably, Spain and Portugal—only started developing environmental regulations during the 1980s, a factor that may have an impact on their relative propensity to import hazardous wastes.

10. Principal-agent problems arise in situations where "a principal (or group of principals) seeks to establish incentives for an agent (or group of agents), who takes decisions that affect the principal, to act in ways that contribute maximally to the principal's own objectives" (Yarrow and Vickers 1988:7). It is used in economic theory to describe situations where the principal and the agent are interdependent (Stiglitz 1987:967): the principal is dependent on the agent to carry out the assigned task, while the agent depends on the principal for remuneration. Two particular characteristics compound the problem. First, the agent

is typically better informed than the principal about her own work effort (there is an asymmetry of information), and second, both parties are assumed to be self-interested and to be maximizing different and contrary objectives. This analysis also carries over well to regulatory situations, where the principals are ultimately beholden to voters, but the agents (firms and bureaucrats) are not. See Weale 1992a:17.

11. See, for example, Weale, O'Riordan, and Kramme 1991, Pridham 1994, and Leonard and Morell 1981.

12. This issue goes right to the heart of a major dilemma facing policymakers trying to implement a strategy of waste reduction. While cheaper or subsidized disposal facilities encourage disposal by the most effective means, they also create disincentives for companies to implement costly waste reduction strategies (Linnerooth and Davis 1987:187).

13. This is a distinction of quality as well as degree. Integrated pollution control (IPC) incorporates a particular regulatory philosophy, whereby the effects of a given activity on the environment as a whole are assessed, as opposed to its effects on a single medium. For example, a decision to close a landfill must also take into account the likely alternative disposal routes for the wastes. This sounds obvious, but in most regulatory systems that sort of decision is made purely with respect to the landfill and its surrounding environment. The point here is that a fully centralized system may not be fully integrated: decisions could still be made without taking into account the effects on the entire environment. For further discussion on IPC, see Irwin 1990 and Skea and Smith 1998.

14. See Weale et al. 1996.

15. There is a sizable literature on the effects of federal vs. unitary systems on environmental policies and, within that literature, many analyses of the effects of different types of federal systems. These themes recur in the case studies and the conclusion, but, for other discussions, see Bowman 1985, Hanf and Toonen 1985, Braden, Folmer, and Ulen 1996, Scheberle 1997, Holland, Morton, and Galigan 1996, Kelemen 1998, and Farrell and Keating 1998.

16. In situations of regulatory capture—where regulatory authorities are effectively controlled by the actors they are supposed to be regulating—the capture of a centralized regulatory agency would be a more valuable prize to the regulated firms than the smaller agencies in more decentralized systems, which are arguably easier or cheaper targets. The extent to which regulatory capture occurs needs to be empirically verified, but where it does there is likely to be an increase in waste importation.

17. Particularly if it is the case, for example, that foreign firms pay more for waste disposal services than domestic firms. Alternatively, it gives disposal firms an additional source of supply.

18. Also on adversarial regulation and regulatory style in the United States, see Kagan 1998.

19. These are: 1) liberal-pluralist vs. statist vs. corporatist styles; 2) active vs. reactive styles; 3) comprehensive vs. fragmented or incremental styles; 4) adversarial vs. consensual vs. paternalistic styles; 5) legalistic vs. pragmatic styles; 6) formal vs. informal network relations (van Waarden 1995:335–336).

20. Similar arguments have been made by Stark (1992) and Steinmo (1989). The former relates government-business-society relations to privatization policies in Eastern Europe, while the latter relates them to tax policies in the United States, Britain, and Sweden.

21. The stronger version of this statement is to argue that such institutional measures actually shape the preferences and organization of societal and business interests. For example in states such as Britain that have a first-past-the-post electoral system, the Green Party has remained extremely marginal in the electoral process (at best gaining representation at local council level); in contrast, the Green Party has a more important political presence in those continental European countries whose electoral systems are based more on proportional representation.

22. Labor groups, another possible group of stakeholders, are omitted because they have not played a key or easily identifiable role in waste trade debates in these cases.

23. On the former, see for example Richardson and Rootes 1995 and Tarrow 1998. On comparative public opinion, see Mertig and Dunlap 1995, Dunlap, Gallup, and Gallup 1993, and Brechin and Kempton 1994.

24. Or, hypothetically, where business interests can outweigh government preferences.

25. There is a powerful counterargument to this statement that accepts that public opinion opposes waste importation but argues that in many democracies power configurations are such that waste disposal facilities are located in impoverished and/or minority communities. These communities are either too isolated from the political process or too much in need of the economic benefits associated with such facilities to voice their opposition on environmental or health grounds (see for example Bullard 1991; United Church 1987; Seager 1993). I discuss this argument in more depth in the empirical studies.

26. The notion of policy communities is often contrasted with that of "issue (or policy) networks"—looser, broader, less integrated configurations of actors lacking the hierarchy of power that characterizes a policy community. See for example Atkinson and Coleman 1989.

27. In his comparison of the strategies adopted and results obtained by antinuclear activist groups in four countries, Kitschelt distinguishes between input structures (role in policymaking process), which may be open or closed, and output structures (policy implementation), which may be strong or weak.

28. For general discussions of implementation and enforcement mechanisms, see Hutter 1999 and 1988, and Jasanoff 1991.

29. The OECD itself is not an international regulatory body, as it has neither supranational legal powers nor financial resources for loans and subsidies. It is, instead, a forum for officials of member countries to meet, exchange information, and coordinate domestic policies with other countries. However, membership implies a commitment to certain forms of policies (free markets, democracy) and openness to scrutiny by other members. In addition, it provides guidelines in many policy areas, including environmental management, and holds extensive collections of statistics on members and nonmembers.

30. See chapter 2 for a discussion of the Basel Convention and related agreements. The final chapter addresses the issue of how OECD nations are likely to react to the banning of most waste exports to less developed countries—for example, whether it will lead to trade diversion, greater self-sufficiency, or, in the worst-case scenario, a boom in the illegal trade.

31. See Gilpin 1987:173–4, for a discussion of the notion of comparative advantage. In addition, another way in which countries may be said to have a comparative advantage in waste disposal may be termed "geological advantage." This occurs when a country has favorable geological or geographic conditions for waste disposal, increasing its assimilative capacity (Strohm 1993:134). The paradigmatic example of geological advantage is Great Britain, which is said to have an extremely favorable geological and climatic structure for the dilution and dispersal (often in the direction of other countries) of all forms of pollution. On the other hand, given the pressures placed upon this system, it is doubtful now how effective natural methods of pollution dispersal still are (see Rose 1990 and Golub 1994). In turn this relates to notions of carrying capacity, where it could be argued that some countries are simply able to manage or safely dispose of wastes through natural or man-made processes. This is a hard measure to assess on a national level with any objectivity.

32. If, however, foreign and domestic wastes are easily substitutable for each other, this would imply greater elasticity in waste disposal capacities and a higher explanatory power for this variable.

33. See World Resources Institute 1987, table 13.5, for an overview of common biological, chemical, and physical treatment processes in North America and Western Europe, and their advantages, disadvantages, and limitations.

34. Note that this conception of costs refers only to immediate costs of disposal borne by the facility operators and reflected in the price charged to waste generators. It does not refer to longer-term costs—often borne by regulatory authorities—such as those incurred in the clean-up of old or contaminated sites, as is the case with Superfund in the United States. Costs of disposal also vary according to the type of waste being handled. Finally, these costs are often the subject of government intervention, either in the form of subsidies on the most effective forms of technology or taxes on the least effective—a subject I deal with more fully in subsequent sections. In addition, costs of disposal may also vary according to the structure of the waste disposal industry—the subject of the next section.

35. For example, according to the OECD, "[f]or the United States, some 1985 data for the state of New Jersey reported disposal cost ranges of $200 to $2600 per tonne. Very roughly, the average cost of disposal by means other than incineration was $500 per tonne while for incineration, the average cost was $1500 per tonne" (OECD 1993b:11). In Europe landfill costs are usually around one-half to one-third the cost per tonne of incineration and chemical processing (OECD 1994b, 1993c; Pflügner and Götze 1994).

36. This argument is based on certain propositions about government preferences in a democracy. For example, elected government officials are assumed to have wider objectives—such as reelection—than making money by providing cheap disposal facilities, and their thinking can also be assumed to be more long-term than most firms', as they will inevitably be responsible for cleaning up problem sites.

37. With respect to illegal trade, unfortunately, the opposite principle applies: illegal wastes almost always go to the countries with the cheapest disposal facilities.

38. The membership of the Organization for Economic Cooperation and Development encompasses the world's advanced free-market democracies. The original OECD member states are: Austria, Belgium, Canada, Denmark, France, Germany, Greece, Iceland, Ireland, Italy, Luxembourg, The Netherlands, Norway, Portugal, Spain, Sweden, Switzerland, Turkey, the U.K., the U.S., Japan, Finland, Australia, and New Zealand. Between 1994 and 1996, Mexico, the Czech Republic, Hungary, Poland, and the Republic of Korea joined the OECD.

39. There is a well-established literature on the differences between Germany—Europe's Green Knight—and Britain—the "Dirty Man" of Europe with respect to environmental issues. See for example Rüdig 1993, Weale 1992a and 1995, Boehmer-Christiansen and Skea 1991, Boehmer-Christiansen and Weidner 1995, Gordon 1994, and Marshall 1998.

40. Ideally this model should be tested using other OECD countries as well. However, the limitations at this time on data availability (in terms of consistent and reliable data on waste importation practices over a number of years and from a number of sources) has limited case selection to the set examined here.

41. On interstate differences and waste transfers, see Lester et al. 1983 and Rabe 1997. U.S. waste exports to and imports from other OECD countries are relatively much smaller, with the exception of Mexico and Canada. See O'Neill 1999.

42. See Buller, Lowe, and Flynn 1993, Héritier et al. 1994, and Vogel 1995 for arguments about harmonization and the EU.

43. For detailed discussion of this problem, see Montgomery 1995.

44. For example, neither Britain nor Germany reports wastes imported or exported for recycling purposes. Belgium includes all wastes in its reported figures. Japan did not even publish data on waste generation and disposal until very recently. Most countries have very different listings of what they consider

to be hazardous wastes; these listings usually include industrial wastes, but may or may not include other categories of wastes, e.g., hospital, agricultural, or toxic household wastes. See Dowling and Linnerooth 1987 and the empirical chapters of this book for further discussion.

45. For general OECD overviews of waste trading policies, see OECD 1985 and 1993b. On waste trade statistics, see OECD 1993a, 1994a, and 1997a. On case-specific environmental performance, see OECD 1993c (Germany), 1993d (Japan), 1994b (Britain), and 1997b (France).

Chapter 4

1. Throughout, the terms Britain, Great Britain, and the United Kingdom are used interchangeably. The difference is that the term "United Kingdom" includes Northern Ireland, while "Britain" refers only to England, Scotland, and Wales.

2. However, official reporting of waste importation figures did not begin until 1988. Waste importation most likely began in the 1970s, at least in part a function of the new laws set up under COPA 1974 (see below). It is also possible to point to several contextual factors, in particular Britain's accession to the EC in 1973. However, as argued later, that Britain joined the EC at this point does not explain why it became a waste importer rather than an exporter.

3. Figures compiled by the British Hazardous Waste Inspectorate and Her Majesty's Inspectorate of Pollution. Note that the 80,000 tonne import figure is in sharp contrast to the one issued by the Department of the Environment. This could be because of as yet unresolved differences between waste definitions; it also could be that the larger figure includes wastes imported for recovery. Whatever the reality, the trend is still sharply apparent.

4. According to British Department of Environment figures (1995), the only waste exports that occurred in the period 1988–1994 were 525 tons of lycra and isocyanate wastes exported to Finland from Northern Ireland between 1991 and 1992. The story for waste export for purposes of recovery is somewhat different. According to the then environment minister Robert Atkins, "exports of hazardous recoverable waste from the U.K. to non-OECD countries amounted to 16,300 tonnes in 1991. The figure was halved in 1992" (ENDS Report 228, January 1994:34). According to the Environmental Data Service (ENDS), most of these wastes were destined for Southeast Asia and the Middle East (ibid.). The most ingenious export scheme noted by ENDS was a plan by a British businessman to export wastes high in cellulose to Hong Kong, where it had been found they prove immensely suitable for growing mushrooms. Alas, the outcome of this scheme was not reported (ENDS Report 127, August 1985:8).

5. Note that these import figures do not include any wastes imported for recycling or reuse: reporting these is not required under British regulations. However, some figures are available: "The DoE's latest annual digest of statistics

suggests at most 3,700 tonnes of the 45,800 tonnes of hazardous waste imported into England and Wales in 1992/3 came in for reprocessing or recycling" (ENDS Report 233, June 1994:28).

6. *Department of the Environment, Transport and the Regions Digest of Environmental Statistics No. 20*, 1998: "under the new European Regulation, international movements of all wastes shipped for disposal are notifiable whereas under the 1988 regulation only hazardous wastes were notifiable."

7. Data on the countries exporting to Britain in earlier years bears these trends out. What is notable is the stark rise in imports from Germany, especially following unification: exporting barely 1000 tonnes in 1990/1, Germany is now the leading waste exporter to Great Britain by a long way (ENDS Report 233, June 1994:28). Other countries that export their wastes to the U.K. include Australia, New Zealand, and Hong Kong—all of which have long-standing Commonwealth links with Great Britain. ENDS Reports note that the number of countries exporting to Britain has increased during this period (ENDS Report 199, August 1991; see also Vallette and Spalding 1990:353).

8. PCBs are ranked high on the OECD's "red list" of hazardous wastes; sulfuric acid, chlorinated solvents, and pesticides are on the "amber list" (OECD 1993b, appendix 2). For the others, more precise detail on waste content is needed.

9. Along with France—see chapter 6.

10. For an overview of Britain's environmental history, see B. W. Clapp 1994.

11. For contrasting explanations of Britain's regulatory system, see Vogel 1986, Moe and Caldwell 1994, and Weale 1992a. What each of these has in common is that Britain's style and structure of environmental regulation is the product of institutional forces of different kinds. Vogel, for instance, discusses the way government-business relations have evolved since the nineteenth century, while Moe and Caldwell relate regulatory practices to the parliamentary system of government.

12. See Skea and Smith 1998, which also covers the differences between Integrated Pollution Control (IPC), an early domestic initiative, and Integrated Pollution Prevention Control (IPPC), a principle embodied in 1996 European legislation and which is the cornerstone of policy harmonization within the EU. This is discussed further in chapter 7.

13. Alternatively, the government moved away from a regulatory role as a provider of goods and services—social regulation—toward a form of economic regulation—protecting consumers from the excesses of private monopolies (Weale 1996:125). A variety of other explanations have been offered as to why this was the case. Some relate it to the growing influence of the EC in domestic affairs (Blowers 1987). Others highlight Thatcher's 1988 speech to the Royal Academy of Sciences, in which she praised the role of British scientists in the discovery of the ozone hole above Antarctica.

14. Britain defines "special wastes" (the term "hazardous waste" is not an official term in the U.K.) to include "any *controlled waste* [any waste disposed of

via landfill or incineration] consisting of or containing substances that are 'dangerous to life'" (OECD 1994b:65). For a discussion of the British system of waste classification, see Dowling and Linnerooth 1987:135–140; subsequent changes as a result of European legislation will be discussed separately. The term "special wastes," as used in Britain, encompasses a narrower range of materials than elsewhere. According to Wilson, "the definition of a special waste is thus framed specifically to isolate only those hazardous wastes potentially dangerous to life if encountered after fly-tipping" (Wilson 1987:241). Various independent commissions—for example, the HWI, the House of Lords Select committee on hazardous wastes, and the Confederation of British Industry—have been highly critical both of the official definition and the haphazard way in which data has been gathered on waste generation. Hildyard notes that if the CBI definition were to be adopted, figures for hazardous waste generation in the U.K. would be in the realm of 12 million tonnes per year, as opposed to the currently reported 2–3 million tonnes (Hildyard 1986:215).

15. The most important piece of conservation/development legislation is the 1947 Town and Country Planning Act, which in effect put all private development under public control (Garner 1996). Although elements of hazardous waste policy—in particular, the designation of landfill sites—fall under the former category (Mädel and Wynne 1987; Yakowitz 1993), it is seen primarily as a pollution control issue in terms of the potentially harmful effects of hazardous waste disposal on the local environment. This division between policy priorities will reemerge in the discussion of Britain's environmental movement.

16. Some relate this comparative lack of concern for air, water, and soil pollution to the fact that Britain's natural assimilative capacity—its ability to absorb pollutants into the ecosystem—is very high compared with most countries. With short, fast-flowing rivers, favorable wind currents, and porous geological structure, the natural ravages of environmental degradation have not made themselves felt quite as soon as they might have, and indeed have contributed to the somewhat laissez-faire attitude adopted by the governing authorities (Weale 1995).

17. Another body—best referred to as a quasi-governmental authority—created to deal with environmental issues is the Royal Commission on Environmental Pollution, established in 1969. Although the commission has no executive power, it plays an important advisory role, particularly in recommending the creation of a unified pollution inspectorate, Her Majesty's Inspectorate of Pollution.

18. With respect to hazardous waste management, the 1990 Act endows a "duty of care" on any person who handles the waste at any stage of its life cycle; this is in contrast to COPA, where only those responsible for final disposal of the waste bore legal liability. In turn, this has also involved the establishment of stricter licensing requirements. Breach of this duty of care is now a criminal rather than civil offense (Cooke and Chapple 1996:12).

19. In Britain responsibility for managing counties and towns rests with local councils, supported by their own administrative agencies. Council members are

elected officeholders and almost always representative of the major political parties. While this basic pattern remains constant, many changes—over and above those arising from changes in funding arrangements—have been and are due to be effected in local government organization. These began in the late-1980s with the abolition of the councils governing the bigger cities—such as the Greater London Council—and continued with the proposed amalgamation/abolition of some county councils.

20. "Die politischen Bühnen auf der subnationalen Ebene—die Local Authorities—sind im Unterschied zu den Ländern in der Bundesrepublik—nicht systematisch institutionell in den Entscheidungsprozeß auf der zentralen Ebene eingebunden" (translation by author). See also Marshall 1998.

21. Until 1990 local councils had relied on a rating system, a tax based on property values, with wide exemptions according to income. This system was severely flawed. However, the proposed alternative—the Community Charge, or poll tax, whereby households were to be taxed a lump-sum per member, almost regardless of income—was manifestly regressive. Thatcher's attempts to push the tax through in the face of strong opposition are widely credited with causing her downfall. The Council Tax, the system that replaced the poll tax, is a mixture of the two, assessed according to property values and number of household residents.

22. For further discussion of these changes, see Williamson 1995 and Rhodes 1994.

23. See also Weale 1996:127–128.

24. Lord Hesketh, quoted in Carter and Lowe 1994:267. Official policy on this was thrown into sharp relief when in 1985 an independent peer, Lord Shannon, put forward a proposal for a centralized waste management policy during a debate in the House of Lords, citing "overwhelming support" for this plan from businesses, local authorities, and environmental groups, thus prompting Lord Hesketh's comments. Hesketh did add that "there is a place for the Government to endorse a national waste strategy as a guide to local authority decisions and a confidence building backdrop for the private sector to make the necessary substantial investments over the next few decades" (cited in ENDS Report 184, May 1985:23). This was the first official mention by central government for the need for a national strategy—one that was not followed up on for another ten years.

25. For a history of the development of the British waste industry, see B. W. Clapp 1994, chapter 9 ("Any Old Iron?"). This detailed account of the industry, from its origins in the activities of lone entrepreneurs to its current incarnation, demonstrates that private enterprise has a long history in Britain and is not just a product of the Thatcher years.

26. Earlier estimates put this figure at 165 (Wilson 1987:238) and "nearly 200" (Mädel and Wynne 1987:233).

27. See also Porter 1998.

28. However, the resources devoted to the HWI pale in comparison to equivalent bodies elsewhere: in 1987 "the U.S. EPA had 20,000 on staff to monitor air pollution, and the Dutch Inspectorate had 1,000, but Britain employed only 32 inspectors" (Moyers 1991:92).

29. For example, the 1990 report of the HMIP criticized many WDAs for concentrating on their "operational activities to the detriment of their regulatory duties" (cited in ENDS Report 187, August 1990:4). This view echoes those found in the first report of the HWI (1985) (as reported by ENDS), which severely criticized the capability of the WDAs in handling hazardous wastes. Among its findings was that the management of hazardous wastes is of relatively low priority relative to that of other sorts of wastes (ENDS Report 125: June 1985). More recently, the WRAs have complained about shortages of revenue (they rely primarily on a charge levied on waste disposal firms) and staff (ENDS Report 254, March 1996:13–14).

30. For further discussion of policy communities and the wider concept of issue networks, see Winter 1996:25–29; Judge 1993; Rhodes 1988, and for the original argument, Richardson and Jordan 1979. On the whole, these works concur that in pollution control policy, the notion of policy community is more useful.

31. One of the main forums for scrutiny and monitoring of legislation occurs in Great Britain via the system of Parliamentary Questions, where Members of Parliament are entitled to pose questions to the Prime Minister or the relevant cabinet Minister about legislation. The technicalities of environmental policy are often beyond the realm of expertise of most members of Parliament.

32. The changing attitudes of the Conservative Party to environmental issues are charted by Flynn and Lowe (1992), who quote extensively a 1985 report written for the party on these issues by Andrew Sullivan. Sullivan argued for an approach that would "correspond with conservative sentiments among ordinary people" (Flynn and Lowe 1992:21), hence focusing on the traditional values of preserving the national heritage, as opposed to adopting a "greener" platform. The Labour Party under the leadership of Tony Blair maintains a rhetorical commitment to environmental issues, but is traditionally antipathetic to them in practice. The Liberal Democrat Party has, despite its small parliamentary delegation, played a substantial role in putting environmental issues on the agenda.

33. More recently, the term "best practicable means" has been superseded by "best practicable environmental option" (BPEO), and under the Environmental Protection Act, firms are required to adopt the "best available techniques not entailing excessive costs" (BATNEEC) (Jordan 1993:414).

34. See Vogel 1986:87–90 for figures and analysis of the lack of prosecution of errant firms. In fact, most of the prosecutions that have occurred have been aimed at smaller firms. He also quotes government representatives, who argue that the cooperative approach—the carrot over the stick—elicits a much more favorable response from industry, and hence a higher degree of compliance with regulations at a lower cost—precisely the opposite of what occurs in the United

States. This practice came under criticism as early as 1972 in an article in *New Scientist*, where Jon Tinker "said Britain's environmental policy was not suited to an advanced industrial society. Offenders [i.e., polluters] 'were taken quietly on one side by the prefects and ticked off for letting the side down. There is no need for prosecution; the shame of being found out is reckoned to be punishment enough'" (from "Britain's Environment: Nanny Knows Best," cited in Allen 1992:219–220).

35. From a report in the *Financial Times* ("Getting out of Holes in the Ground"), July 13, 1996.

36. From Barclays de Zoete Wedd Report, released by the ICC Information Group, April 1, 1993 (ICC Report No. 048401). According to this report, shares of a market valued at £2.5–£3 billion are as follows: Leigh: 4%; U.K. Waste: 3%; Cleanaway: 4%; Shanks/ReChem: 5%; Biffa: 4%; other major players: 8%; local authorities and small operators: 75%. Note that these figures cover the entire waste disposal industry; the major players are the ones involved in hazardous waste disposal. However, these figures do give an indication of how low the level of industry concentration is in the U.K. in comparison with its European neighbors, where the top five players in each country have a much higher share of the market (ENDS Report 248, September 1995:14).

37. The internationalization of Britain's waste industry coincides with the increase in waste importation. In terms of hazardous waste disposal, the British firm Waste Management International has historically operated plants in nineteen countries.

38. ENDS Report 233, June 1994:28. By way of comparison, the average disposal cost per tonne of hazardous wastes in Britain is about £50 (see discussion below).

39. House of Lords Report 272-1 (Report), session 1980–81, 38, cited in Montgomery 1992:208.

40. Government-waste industry relations are not always harmonious. The waste disposal industry has criticized government actions on a number of grounds, including its refusal to centralize waste regulation, which would create a "one-stop shop" for industry, and its slow enactment of policy promises. As one analyst put it, "British players have suffered, not from legislative overkill, but by the void between legislation and its timely, orderly and effective enforcement" (Dr. David Owen, leading waste industry analyst, cited in ENDS Report 248, September 1995:14).

41. Chapman et al. (1997) point out this story as one of the catalysts for the emergence of "green" journalism in the U.K., marking the start of "the most recent period of sustained media interest in environmental issues" (ibid.:43).

42. From *The Environment in Your Pocket 1998*, published by DETR. See http://www.environment.detr.gov.uk/des20/pocket/env52.htm. These levels of concern are comparable with Germany, as a much earlier study showed, in Heidenheimer, Heclo, and Adams 1990:317, citing a 1982 survey carried out

by the Commission of the European Communities (*The Europeans and Their Environment*, Brussels, 1983). See also Norris 1997.

43. For example the National Trust, whose remit is the protection of buildings and areas of historical or natural interest, the Royal Society for the Protection of Birds, and the Rambler's Association represent important rural constituencies—a group that the Conservative Party (or any ruling party) has an interest in courting (Blowers 1987).

44. See Norris 1997 on the distinction between "old" and "new" Green politics in Britain.

45. Garner reports membership figures in 1989 for FoE at 120,000 and for Greenpeace at 320,000 (Garner 1996:64; Grant 1995:14). Membership figures for 1995 for FoE and Greenpeace respectively are 180,000 and 279,000 (Rawcliffe 1998:23).

46. Or, put another way, FoE has moved "into the suburbs of British politics" (Boehmer-Christiansen and Skea 1991:81).

47. This included a somewhat unlikely alliance between the "New Age Traveler" types and the "Green Wellie Brigade." Television news bulletins at the time of the motorway protests in the summer of 1995 were full of reports of country ladies serving tea and sandwiches to the youngsters in the trees. History repeated itself in May 1997 during the protests over the second runway proposed for Manchester airport.

48. See also the 1998 DETR study cited above, which shows high public concern across a range of environmental issues.

49. See "Environmental Policy Must Take into Account Public Opinion," *British Medical Journal*, October 17, 1998.

50. On the fluctuating fortunes of Britain's Green Party, which reached a peak in the 1989 European elections with 15% of the popular vote, see Rootes 1995. The Greens have had the most success at the local level, securing as many as twelve seats nationwide. However, on average they are only able to contest around 10% of council seats.

51. From International Environment Reporter (BNA), December 11, 1988 (cited in Montgomery 1992:218).

52. Department of the Environment, *United Kingdom Management Plan for Exports and Imports of Waste*, London: HMSO, 1996. In general, under this plan, imports for disposal are banned while imports for genuine recovery purposes are permitted. However, there are several notable exceptions: imports for disposal may be permitted when the exporting country does not have adequate facilities for "environmentally sound management" of the wastes—this rule applies mostly to LDCs. More interesting are the number of loopholes for EU members (non-EU industrialized countries are treated much more strictly). Eire and Portugal are allowed to import wastes to the U.K. for an indefinite period; imports from other EU members, rather than being stopped outright, are to be phased out gradually over three years, starting from a high baseline (ENDS

Report 256, May 1996:33–4). While these exceptions apply only to wastes imported for incineration, it should be noted that the ash resulting from these processes is in fact landfilled; the significance of this fact will become clearer in subsequent sections. It is also the case that the government has decided not to draw up a list of wastes to be proscribed for importation purposes—again, leading to accusations of issue-fudging.

53. The article in ENDS Report 256, May 1996 leads with "Lobby by incineration firms pays off in U.K. waste trade plan" (p. 33); earlier reports (241, February 1995 and 233, June 1994) trace this process.

54. This might be the only advantage that the industry sees in a decentralized system. In order to open and operate a waste disposal facility, firms have to deal with a whole network of agencies, including not only the WDAs but also the local landuse planning agencies; the waste disposal industry has clearly indicated that it would prefer a "one-stop shop" (Carter and Lowe 1994:269).

55. ECOTEC, a British environmental consulting firm, has estimated the cost of cleaning up the U.K.'s stock of contaminated land at £10–30 billion, and rising. At 1989 levels of spending, furthermore, the government was investing only 19% of the sum required to achieve the cleanup goals set for the year 2000 (ENDS Report 201, October 1991:4–5).

56. This figure is supported by Hildyard: "According to the Hazardous Waste Inspectorate, there are 4,202 landfills in England and Wales, of which an estimated 1,145 are licensed to take hazardous wastes. In addition to those landfills, there are some 737 land based disposal facilities, including 99 storage sites, 53 lagoons, 5 "reception" pits and 14 mine shafts" (Hildyard 1986:225).

57. This includes a very small proportion—about 454 tonnes—that went for recycling or recovery.

58. Report issued by Barclays de Zoete Wedd, released by the ICC Information Group, April 1, 1993 (ICC Report No. 048401).

59. The glowing terms used by the government to describe the quality of its landfills actually obscure the debate going on over the safety and sustainability of such practices, which have in recent years come increasingly under the scrutiny of the EU, which seeks to impose a ban on codisposal. The British government has lobbied vigorously against such efforts, but is fighting a losing battle (ENDS Report 234, July 1994:30). Finding alternative disposal routes for wastes currently codisposed, it is estimated, "would add about £160 million to industry's annual waste disposal bill" (ibid.:31).

60. "U.K. Waste Criticizes Landfill 'Double Standards'," *Haznews* 83 (February 1995):1.

61. Allen (1992:171) argues that British licensed incinerators would actually be illegal in other West European countries.

62. Ironically the BSE (or "Mad Cow Disease") scare of 1996 might well have positive spinoffs for waste incineration: if Britain carries out its plans to slaughter several million cattle over the next few years, new facilities will be

needed—possibly courtesy of the EU. However, and this is by way of an aside, according to ENDS, "regulations exempting huge quantities of bovine protein and tallow from waste management licensing were slopped out by the Government in mid-May [1996]" (ENDS Report 256, May 1996:34).

63. NAWDC representative, testifying before the House of Lords Select Committee on Science and Technology, 1981 (cited in Montgomery 1992:206).

64. See also Barnes 1997:182–183 for cost breakdowns.

65. Department of the Environment, *Waste Management Planning: Principles and Practice*, London: HMSO, 1995. See also ENDS Report 234, July 1994 and *Haznews* Number 104, November 1996.

66. Reported in the *Daily Telegraph*, March 22, 1995. In its final form, the tax, implemented in October 1996, is levied at £7 per tonne (£2 for relatively inactive wastes); the government estimated that it would raise around £450 million a year in revenue (ENDS Report 258, July 1996). Problems with the tax are discussed in "Working Under Blindfold on the Landfill Tax Review," ENDS Report 276, January 1998:20–23).

67. "Agency Makes a Mess of Waste," ENDS Report 280, May 1998:25–28.

Chapter 5

1. According to West German government documents, in 1986 West Germany exported a total of 943,528 tonnes of hazardous wastes, of which 653,000 tonnes went to the DDR. In 1988 it exported 1,058,067 tonnes, 684,306 to the DDR (Deutscher Bundestag, Drucksache 11/2075 (West German Federal Parliament Document) and statistics of the Hanseatisches Baustoffkontor in Lübeck; cited in Vallette and Spalding 1990:372–73).

2. The Schönberg waste dump, located near Lübeck, just over the border between the two Germanys, has become emblematic of the environmental degradation that occurred in East Germany under the communist regime, as well as the carelessness with which the West Germans treated their neighbors in this regard. For a discussion of the waste management practices of the East German regime, see Vallette and Spalding 1990:41–42. Currently, the West German Länder that used to export to Schönberg—for example, Hamburg—are making arrangements to ship wastes elsewhere (within the country), while existing facilities are being brought up to standard and new ones built.

3. In 1990/1 Germany exported 987 tonnes of waste to Great Britain. In 1991/2 it exported 3,190 tonnes, and in 1992/3, 9,198 tonnes (ENDS Report 233, June 1994:28). See also *Haznews*, December 1992, which notes that while exports to France decreased between 1988 and 1990, there were large increases in the amount of waste shipped to the Netherlands, Belgium, and Denmark.

4. For example, in November 1996 German authorities became embroiled in a dispute with Lebanese authorities when it was discovered that thirty-six containers of hazardous wastes originating in Baden-Württemberg had been smuggled into Lebanon.

5. It is estimated that between 1985 and 1990 roughly 8% of Western Europe's hazardous wastes were incinerated and 70% "delivered to landfills" (Stanners and Bourdeau 1995).

6. In turn, government figures released in 1992 indicate that roughly two-thirds of the total hazardous wastes generated in the FRG are sent to offsite disposal facilities (*Haznews*, December 1991:18).

7. Disposal on or into the North Sea was phased out at the end of 1989. This had been one of the major disposal routes for German hazardous wastes (Sierig 1987, OECD 1993c).

8. For discussions of the integrated treatment facilities in Bavaria and Hessen, see Linnerooth and Davis 1987; Linnerooth and Kneese 1989; Defregger 1983, and Beecher and Rappaport 1990. For discussion of in-house treatment by German firms, see Beecher and Rappaport 1990. For example, they cite BASF, which "operates one of the world's largest chemical plants, stretching 5 km along the Rhine at Ludwigshafen and employing 50,000 workers ... the plant has seven large incinerators with an annual capacity of 135,000 metric tons. [This] probably exceeds the total commercial rotary kiln incineration capacity in the United States" (Beecher and Rappaport 1990:34). On Baden-Württemberg, see Barta 1997.

9. Efforts are also being made to bring the former states of the DDR in line with standards prevailing in the older Länder. For example, in 1991 roughly DM 50 million were spent on improving the Schönberg site, and another DM 70 million spent in 1992 (OECD 1993c:57). The process of cleaning up old sites and improving those still in use in the new Länder is, however, proving costly and slow.

10. *BNA International Environment Reporter*, February 7, 1996.

11. For reasons why there is generally a higher level of environmental threat perception in Germany than in Britain, see Boehmer-Christiansen and Skea 1991 and Héritier et al. 1994.

12. Median disposal costs also vary by Land. According to a report in *Umwelt* (November/December, 1995), they vary from DM 750 per tonne in North Rhine-Westfalia to DM 4,000 per tonne in Thüringen.

13. See also chapter 4 for comparative data on the United Kingdom.

14. For discussion of Germany's role in different sets of international regime negotiations, see Porter and Brown 1995, Weale 1992a, Boehmer-Christiansen and Skea 1991, and Levy 1993.

15. For discussions of Germany's federal division of powers and resulting relations and issues, see Peters 1992 and Bulmer 1989. Not only does the constitution contain mechanisms that determine the allocation of revenues between central government and the Länder, but it also contains mechanisms for reallocation of funds from the wealthier to the poorer Länder.

16. For example, the Prussian General Trade Ordinance of 1845 placed controls on industrial emissions leading to air pollution (Weale 1992b:161).

17. See also Rose-Ackerman 1995:45–47.

18. "Bedingt durch diesen föderalistischen Staatsaufbau, der sowohl Elemente der Kooperation als auch der Konkurrenz enthält, existiert in der Bundesrepublik eine Vielzahl von politischen Ebene, die sowohl weitgehend unabhängig voneinander agieren, als auch auf vielfältige Weise miteinander verflochten sind. Im Bereich der Luftreinhaltung sind vertikale wie horizontale Formen der Politikverflechtung auf Regierungsebene (Umweltministerkonferenz), auf Parlamentsebene (Umweltkabinett), auf administrativer Ebene ... sowie auf parteipolitischer Ebene ... von Bedeutung" (Héritier et al. 1994:52–3; translation by author).

19. With the exception of military wastes, which are in the province of the Defense Ministry.

20. The definitions of hazardous, or special, waste ("Sonderabfall") in Germany have undergone considerable evolution since the term was first introduced. Two definitions existed side by side in the original 1972 legislation. The first was a more technical term, defining the properties of wastes considered hazardous; the second, a more administrative term, defined "excludable wastes" as those that should be disposed of separately from ordinary household wastes. Over the following five years, the federal government worked to produce a list of 86 hazardous substances that would form the minimum standardized list for Land regulations. Subsequently some Länder, notably Hessen, have expanded this list considerably, and in 1990 a newly expanded list covering a total of 332 waste codes was introduced (Dowling and Linnerooth 1987:125–132; Czech 1996). Efforts are currently underway to harmonize state practices both nationally and in accordance with international agreements. Most significantly, one of the main weaknesses identified with German hazardous waste classification—the exclusion of "residuals"—is being dealt with under the Kreislaufwirtschaft legislation, to be introduced in 1996. Under this legislation this distinction is no longer maintained, although wastes for reuse (formerly categorized as an economic good, and hence under no notification requirements) and for disposal are separated (Czech 1996). One of the main differences between the German system of waste classification and those of its European counterparts is that a waste that contains any of one of the listed hazardous substances is classified as hazardous. Other countries (U.K. and the Netherlands, for example) define a waste as hazardous only when a certain concentration level of the substance concerned has been reached (Wynne 1987).

21. For a discussion of the role of local governments in Germany, see Peters 1992.

22. Integrated or comprehensive facilities being those that utilize a variety of disposal techniques, usually incineration and physical/chemical reprocessing.

23. North Rhine-Westfalia is in fact Germany's largest waste producer, "generating nearly twice as much hazardous wastes as the other [western] Länder combined" (Linnerooth and Davis 1987:176).

24. Bavaria, Hessen, and North Rhine-Westfalia are common cases for the study of waste disposal practices in Germany. In addition to Linnerooth and Davis 1987, see also Defregger 1983 and Linnerooth and Kneese 1989.

25. For example, in 1992 the BMU earmarked DM 3990 million, over and above its regular budget, for "loans targeted at environmental-protection measures in the fields of air-pollution reduction, waste-water treatment, rational use of energy, waste avoidance and waste disposal" (Randlesome 1994:169). See also "Approval for Schleswig-Holstein Special Waste Firm," *Haznews* 68, November 1993; "Baden-Württemberg Special Waste Rises in '91," *Haznews* 59, February 1993; and "Länderregelungen für die Sonderabfall-Entsorgung," *Umwelt* 25:10/11, 1995. This last analysis points out that Land organizations in most states (with the exception of North Rhine-Westfalia) play a lead role in regulating the waste disposal market, noting that "these organizations cover costs, but are not profit oriented" (ibid.:513; translated).

26. For a similar comparison, see Marshall 1998.

27. In Kitschelt's characterization, West Germany has a closed input structure, based on its "centripetal party system, organized along class and religious cleavages, weak legislature and inaccessible executive" (ibid.:66), but a weak output structure, owing in particular to the federal system and the independent judiciary. This, he argues, affected actions of the antinuclear movement, which responded to the opportunity structure it faced by adopting confrontational strategies at the implementation stage.

28. The industry actors are, on the whole, not individual firms but rather trade, or "peak," associations. For technical/scientific advice, the government relies on established bodies of experts.

29. Katzenstein refers to the arrangements prevailing in Germany as "democratic corporatism," distinguished by three traits: "an ideology of social partnership expressed at the national level; a relatively centralized and concentrated system of interest groups; and voluntary and informal coordination of conflicting objectives through continuous political bargaining between interest groups, state bureaucracies and political parties" (Katzenstein 1985:32). In contrast, the British system is more often characterized as pluralist.

30. In addition to their superior resources and informal contacts with policymakers, other reasons are cited for the dominance of industry in the formation of environmental policy. First, the established tradition in Germany is for environmental policy to be transmitted "von oben nach unten," or top-down. Environmental pressure groups, with their roots in community organizations, and activist groups are often thought to wield a distorting rather than a corrective influence on government policy (Héritier et al. 1994:64–5; citing an interview with the EURES Institute—a scientific body—August 1993).

31. This, according to Rose-Ackerman (1995:71), is one of the fundamental weaknesses of the German system of environmental regulation: a tendency to see environmental problems as technological (hence requiring technocratic solutions). She argues that the exclusion of private environmental groups from the formal process of policymaking, as well as the lack of openness in decision-making procedures of many of the expert groups brought in as consultants, reinforces this tendency.

32. According to Weale, the groups most commonly called upon for their advice are environmental lawyers and engineers (Weale 1992b:178).

33. A further difference is that the British tend to prefer "generalists" rather than specialists when it comes to setting policy goals and practices (Enloe 1975; Boehmer-Christiansen and Skea 1991:114).

34. The CDU, or Christian Democratic Union, along with its Bavarian counterpart the Christian Social Union, is Germany's largest party. Its electoral platform is pretty much center-right. The SPD, or Social Democratic Party, has a more leftist position, drawing much of its support from union members. The FDP, or Free Democrats, is situated ideologically between the CDU and the SPD. From 1982 to 1998 the electoral mandate has rested with the CDU-CSU under the leadership of Chancellor Helmut Kohl. In October 1998, federal elections led to the succession of an uneasy coalition between the SPD and the Greens (see below).

35. "Insbesondere regionale Probleme ... haben auf dieser Weise größere Chancen, auf die politische Agenda zu gelangen und bürgernaher behandelt zu werden als in zentral regierten Ländern, deren Kommunen wie in Großbritannien in ihren Handlungsmöglichkeiten stark vom Zentralstaat abhängig sind" (Héritier et al. 1994:52; translation by author). Although Héritier et al. generally favor this system of government, they do point out its disadvantages, notably, a tendency to "pass the buck" when taking responsibility for policy decisions, and the possibility for conflict—and hence deadlock—between the Bund and the Länder (ibid.:53).

36. See Rose-Ackerman 1995, chapter 2. In comparing the U.S. and German legal-administrative systems with respect to environmental policy, she finds the German system to be a good deal less open than that in the United States. For example, the (automatic) role of the courts in monitoring and restraining the administration is much more significant in the U.S., while in Germany, it is more the case that the courts exist protect individual rights against administrative excess.

37. For a discussion of the role played by the federal court in the interpretation of the precautionary principle, see Weale 1992b:171. For example, the courts have ruled that risk (such as that entailed in the construction of nuclear power plants) is on a continuum with actual environmental degradation, hence extending the reach of the principle.

38. For discussions of Germany's broader regulatory culture, in which its system of environmental regulation is embedded, see, for example, van Waarden 1995, Dyson 1992, and Richardson 1982. Dyson argues that "regulation [in the FRG] is so deeply engrained as a social and political phenomenon that it is possible to identify one of Germany's distinctive characteristics as its 'regulatory culture'" (Dyson 1992:1). He argues that German regulation is driven by "the preoccupation with order" (*Ordnungspolitik*), and by legalism, "which enshrines the primacy of legal rights and procedure" (ibid.:10). The hallmarks of German regulation are, therefore, predictability, reliability, and "public-regarding" behavior.

39. "Dieses Prinzip besagt, daß dem Entstehen schädlicher Umwelteinwirkungen vorbezeugen ist und bereits auf der Konstruktions—und Produktionsebene umweltfreundliche Varianten zu bevorzugen sind." I discuss the second part of this principle—the introduction and use of environmentally-friendly technology in production processes—in more depth later on.

40. See for example Rehbinder 1988 for a review of possible different interpretations of the principle.

41. Jordan and O'Riordan also discuss the role the precautionary principle has played in forcing technological development by German firms and the ways it has affected international policies and the policies of other countries.

42. Violation of hazardous waste disposal regulations—a civil offense in Germany—results in the imposition of heavy fines on the offending firm (Linnerooth and Davis 1987:185). Germany is also one of the few countries to operate a "Green Police" force. Operating near Frankfurt, this small unit of about 10 officers reacts to complaints from the public and are responsible for monitoring dangerous or hazardous loads in transit (Randlesome 1994:173).

43. Brickman, Jasanoff, and Ilgen (1985) argue that the German system, with its emphasis on uniform standards and policy frameworks, resembles the United States more than Britain and France. On the other hand, its formalized system of interest representation, whereby the participation of certain groups is actually laid down by law, puts it closer to the more centralized systems of its European counterparts.

44. For a discussion of the various nonparty environmental organizations in the Federal Republic and the contours of recent German environmentalism, see Blühdorn 1995. He identifies the most important as the German League for the Protection of Nature (NABU), the World Wide Fund for Nature Germany, the German Alliance for Environment and Nature Conservation (BUND), Greenpeace Germany, and Robin Wood (Blühdorn 1995:171).

45. See also Wimmer and Wahl 1995, whose figures tell much the same story. Based on more recent data, they show that waste disposal now ranks number two on the list of environmental concerns, following air pollution. More broadly, they also show that environmental issues rank second or third on average in lists of citizens' urgent concerns (always following unemployment; in recent years, environmental concerns have been overtaken by immigration issues (Wimmer and Wahl 1995:30).

46. The study cited is a 1982 survey carried out by the Commission of the European Communities (*The Europeans and Their Environment*, Brussels, 1983).

47. As reported in *The Week in Germany* (German Information Center), March 29, 1991.

48. This electoral success continued through the 1987 Bundestag elections, when the Greens polled 8.3% of the national vote (Héritier et al. 1994:56, table 5).

49. Héritier et al. refer to this as the "*Ökologisierung*" of party platforms, and describe this primarily as an attempt to appeal to younger voters. The SPD in particular, as the party whose constituency was most undercut by the success of die Grünen, is noted for its changing environmental platform (Schmidt 1996).

50. Absent from the analysis so far is the third member of the corporatist triad: labor organizations. It has previously been the case that labor organizations have not played as great a role in the formation of environmental policy as other groups, and, indeed, until the late-1980s trade unions remained reasonably hostile to the notion of extended environmental protection. On the other hand, the main German trade union peak association—the DGB—has been active in producing reports on environmental policy (notably concerning worker health and safety issues) and has advocated greater clarity in decision-making processes. Furthermore, recent analysis suggests that given the economic health of the environmental sector in terms of job creation, trade unions are coming around to a position more pro-environmentalist than previously. For further analysis, see Siegmann 1985, Heuter 1995, and Hildebrandt 1995.

51. See, for example, Katzenstein 1985, Hollingsworth, Schmitter, and Streeck 1994, and Randlesome 1994.

52. The actual status of laws governing inter-Land waste transfers is ambiguous. While some sources (OECD 1993c:57) claim that such transfers are not allowed, others (Linnerooth and Davis 1987:156) claim that the Länder are indeed permitted to export or import wastes from other parts of the country, and it is up to the individual Länder to set their own restrictions.

53. According to this study, which deals with the waste trade among EU nations, some German Länder import small quantities of wastes from neighboring regions in France.

54. These safeguards are designed not least to counteract the possibility of the state abusing its power, a further legacy of World War II. Furthermore, as van Waarden argues, the formal incorporation of societal interests into the policy process—corporatism—differentiates the German system from that of France, for example, which he characterizes as "étatist" (van Waarden 1995:338–339).

55. Boehmer-Christiansen and Skea (1991) argue that while the term "state" is more commonly used than "government" in Germany, in Britain the opposite prevails (ibid.:94).

56. Grant and Paterson 1994:148.

57. With the obvious exception of the Eastern Länder, which still accept large quantities of hazardous wastes from the West. However, it can be argued that these constitute a separate case, in that disposal laws and standards are still being developed in these regions; on the other hand, it is also an indicator of the institutional legacy of the former regime, and the former relationship between the two Germanys.

58. See Pflügner and Götze 1994:53.

59. In addition, while the Germans had achieved considerable success by the mid to late 1980s in reducing emissions of air pollutants, similar patterns of success had not been seen in reducing the production of industrial waste (Wallace 1995:72–3).

60. For a full discussion of the measures contained within the *Kreislaufwirtschaft* in its final version, see Stede 1996 and "New German Waste Law in Force," *Haznews* 104, November 1996:12. The passage of this legislation was by no means smooth: it faced opposition both from industry and in the Bundesrat. Even now, some of the detailed prescriptions remain vague and hence open to different interpretations (Stede 1996). However, even with additional extended deadlines allowed for certain firms, the law came into force in January 1999.

61. See Koss, Malorny, and Stahlke 1994.

62. As reported by the Associated Press, October 6, 1996, and the *BNA International Environment Reporter*, August 21, 1996.

63. In addition, several Länder have chosen to adopt broad cooperative strategies in attempting to address waste disposal problems. Meereis and Hartmann (1994), for example, discuss the way that the environmental ministry in Baden-Württemberg called together representatives from the political parties, firms, waste disposal firms, environmental groups, and other societal groups to hammer out a consensus on strategies for reducing waste exports.

64. Rethmann, writing from the point of view of a waste management expert from North Rhine-Westfalia, argues for the application of free market principles to the waste disposal industry, including opening up more to European competition. These trends are beginning to become more apparent (see Brusco, Bertossi, and Cottica 1996 and Pflügner and Götze 1994) as new varieties of government-private sector cooperation begin to make an impact on the sector. Note, however, that opening up the market could also imply trading in waste disposal technology, where German firms have a noted international edge.

65. On the causes and consequences of environmental degradation in the former DDR, see for example Jancar-Webster 1998 and Kabala 1991.

66. For example, the German environmental movement adapted its strategies and styles to the prevailing political system, rather than vice versa. See Blühdorn 1995.

67. See Abromeit 1990 and The German Council of Environmental Advisors, Environmental Report, 1998 (available in English at www.umweltrat.de/gut98en3.htm) on the extent of voluntary agreements. In response to the Packaging Ordinance, companies banded together to form the Duales System Deutschland, a nonprofit waste management company that collects and returns packaging wastes to contributing firms. The symbol of membership in this organization is the Green Dot (Randlesome 1994:178). While this measure caused great controversy at the EU level, it demonstrates the desire of German industry overall to comply with, rather than contest, state-imposed regulatory

standards and conditions. For further discussion of the effects the Packaging Ordinance had at the European level, see Vogel 1995:88–93.

68. See for example Weale 1992a and Jordan and O'Riordan 1995, both of which discuss the interaction among the application of the precautionary principle, innovation in the environmental technology sector, and the resulting attitude change on the part of societal actors toward the economics/environment tradeoff.

Chapter 6

1. See chapter 3, on case selection and data availability.

2. In addition, according to a 1992 report by a member of the Conseil Économique et Social, "imports of dangerous or toxic wastes, coming from the Federal Republic of Germany (55%), Belgium (27%), Italy, Switzerland and Holland, were 244,000 tonnes in 1988, of which 90,000 tonnes were incinerated and 122,000 landfilled (mises en décharge), and 234,000 tonnes in 1989" (Pecqueur 1992:25; translation by author). See also Pflügner and Götze 1994.

3. See also Buller 1998 for a comparison of the environmental regulation and political structures in France and Britain.

4. See, for example, Hall 1986, Katzenstein 1978, and Zysman 1983. These works discuss French economic policy in changing international and domestic contexts.

5. See Mény 1996:122–126, Hall, Hayward, and Machin 1994, and Stevens 1996. The powers of the President accrue through a combination of those accorded to a head of state in presidential systems and the powers of prerogative under a parliamentary system; notable among these is the President's role in appointing the Prime Minister from the winning coalition in Parliament.

6. The first head of the environment ministry resigned after three years in office, declaring, famously, that it was "the ministry of the impossible" (Prendiville 1994:9).

7. At the regional level, the agencies now responsible for environmental protection are the twenty-four regional directorates of industry, research, and the environment (DRIREs), under whose auspices hazardous waste management now falls (OECD 1997b:43). For a discussion of their role and some of the complexities and weaknesses of their structure and responsibilities, see Bodiguel and Buller 1994:97–99.

8. Of those, 2 million tonnes are classified as "toxic." There was no breakdown given of the relative toxicity of agricultural wastes, of which 400 million tonnes are produced annually.

9. Générale des Eaux is currently the largest European provider of waste management services and the fifth largest global provider. It owns installations

across France and the U.K., operates urban waste management schemes in Germany, Spain, Colombia, Chile, Australia, and New Zealand, and is extremely diversified. SITA operates for the most part in the U.K. and Spain, though it is considering moving into the U.S. market (company reports on Générale des Eaux by SBC Warburg, May 7, 1996, and on SITA by Merrill Lynch Capital Markets, September 17, 1996).

10. According to EU figures, the average cost of incineration per tonne in France is $60; for landfill, $20. This is higher than Great Britain ($40 and $15 respectively), but much lower than Germany ($130 and $55) (Stanners and Bourdeau 1995:578). However, according to *Haznews*, "the average price of disposal in a class 1 landfill has risen from FF350 ($66) per tonne in 1990 to FF750 ($135) per tonne in 1994" ("Hazwaste Management Strategy in France," *Haznews* 93, December 1995:19).

11. In his examination of political opportunity structures in the 1980s, Kitschelt (1986) characterizes France as the "strongest" state due to its closed opportunity structure on the input side and very strong state powers of policy implementation. To an extent, as mentioned above, this situation is currently changing.

12. For example, the ability to call on the Constitutional Council is limited to the President, Prime Minister, and certain members of the judiciary. Courts at lower levels, furthermore, are severely hampered by a large caseload and slow mode of operation. For further discussion, see Mény 1996 and Héritier et al. 1994:134–5.

13. Les Verts, which grew out of protests over nuclear issues, have had a "history of intermittent minor successes" (Cole and Doherty 1995:45). In 1989 (their year of greatest success), they won 1,369 municipal seats and 9 seats in the European Parliament (Prendiville 1994:47). However, their advancement to a position of influence in national politics has been blocked by the electoral system at the parliamentary level, and by the emergence of a rival environmentalist party, Génération Écologie, led by the charismatic Brice Lalonde, environment minister under Mitterrand between 1989 and 1992. Efforts to form an alliance between these two parties ultimately failed. See Faucher and Doherty 1996 for further discussion.

14. For further discussion of French popular opinion and the environmental movement in France, see Shull 1996 and Prendiville 1994.

15. For a fuller account of this controversy, see chapter 5.

16. Another possibility is that government administrators may, unlike their British and German equivalents, favor waste imports on economic grounds. Although a longer study would be able to address this question in more depth, it still remains clear that the central government remains relatively impervious to societal demands on this issue, and that local governments still seem to lack powers to adequately control waste importation into their jurisdictions.

17. See also "Hazwaste Management Strategy in France" (*Haznews* 93, December 1995:17–19). Litvan argues that these measures are too rigid and expensive, preferring instead the British approach of setting a landfill tax at levels high enough to encourage rerouting of wastes for disposal.

18. Most notably, in 1992 it became engaged in what the journal *New Scientist* referred to as "toxic ping-pong." A shipment of Australian wastes containing 18 tonnes of PCBs was bounced from port to port in Europe before being taken in by the French for final disposal. See "Waste Not ... Incinerate Not: Australia's Waste Problems," *New Scientist*, November 7, 1992 (no byline); "France Relents on PCB Shipment," *Chemistry and Industry*, September 21, 1992; and Anderson 1992a.

19. The term "intractable wastes" is used in Australia to refer to a special class of hazardous wastes whose main defining characteristic is that they cannot be disposed of in existing facilities. The term is also used to apply to wastes over which both commonwealth and state governments have jurisdiction—as opposed to other sorts of hazardous wastes that are the responsibility of state governments (McDonell 1991, footnote 3, p. 38). Halon and CFC stockpiles are the most recent additions to this list of state responsibilities. See McDonell 1991:34, and Beder 1991.

20. Australia also exports some wastes for recycling to destinations in Southeast Asia; these experts are banned under a national moratorium as of 1998.

21. As of 1991, 11,000 tonnes of intractable wastes were being stored in New South Wales. Although the bulk of it originated locally, some 15% had been imported—probably from the other states, although this is not clear. Adding other sorts of hazardous wastes would increase this stockpile at a rate of 12,000 tonnes per year. It was also estimated that roughly 10,000 tonnes of halons and CFCs would be discarded over the following ten years. See Beder 1991.

22. For an overview of the history of economic development and environmental protection in Australia, see Walker 1994.

23. The term "commonwealth" applies to the Australian federal government and is not to be confused with the British Commonwealth—the loosely tied group of states sharing a British colonial heritage.

24. There are few extreme differences between the states in terms of their regulatory systems. This is probably due to institutional mimicry: the state of Victoria was the first to adopt an EPA-type structure, and the rest of the states soon followed. See Anton, Kohout, and Pain 1993 for a discussion of some of the problems associated with this non-nationally coordinated system.

25. The first major commonwealth legislation on the environment was the 1974 Environment Protection (Impact of Proposals) Act, whose purpose was less to provide the basis for federal environmental legislation than to monitor and assess the actions of the individual states in this area. At about the same time,

the Whitlam government also established the Australian Heritage Commission and the National Parks and Wildlife Commission.

26. With the exception of the management of fisheries beyond coastal waters and quarantine laws, which have always been commonwealth responsibilities (Saunders 1996:57).

27. See Saunders 1996 for further discussion of the role of the judiciary in Australia.

28. Australia's federal environment ministry was first established in 1971 as the Department of Environment, Aborigines and the Arts. Since then it has gone through numerous permutations as different portfolios—including housing, arts, heritage, and tourism—have been added or taken away (Doyle and Kellow 1995:153).

29. Analysts disagree over the ultimate effectiveness of these cooperative solutions. Kellow (1996) argues that these developments reflect a positive change in Australian federalism, but others such as Christoff (1994) and Papadakis and Moore (1994) argue that these bodies have been relatively ineffectual and have failed to outlast the life of the government that set them up. Some contend that Australia needs a federal environmental protection authority equipped with enforcement powers, if it is to fulfill its obligations under international environmental treaties (Anton, Kohout, and Pain 1993).

30. Although the federal allocation of powers in Australia shares certain similarities with that in Germany, it also differs in significant ways. The most obvious are the absence of the "template model" of environmental legislation and, as a result, once the transboundary nature of environmental problems became more apparent, the emergence of conflict in state-center relations over environmental issues. The federal systems of environmental regulation are further discussed and compared in the introduction to Holland, Morton, and Galigan 1996. See chapter 3 here for further discussion.

31. Mode of policy implementation is not included separately in the discussion here. Briefly, however, with central government control of waste imports and exports—and consequent reduction of industry freedom of action—and fairly strict standard application, it is best, though not ideally, characterized as rigid in the Australian case.

32. The judiciary—important for citizen access in the German system—is not a main channel for public action on environmental issues, as citizens cannot bring cases before it. On the other hand, it has been very supportive of commonwealth efforts to expand its environmental powers (as, for example, in the Franklin River Dam case). For relevant discussion, see Lynch and Galligan 1996.

33. Australia's other main party is the conservative Liberal Party, which most often runs in coalition with the National Party (which represents conservative rural interests). The Liberal Party is noted for its tendency to oppose the expansion of regulatory responsibilities into environmental protection and to favor

developmental interests over environmental ones—although its manifesto for the 1996 general elections proclaimed otherwise.

34. Australia also works on a three-year general election cycle: this also has the effect of keeping the ruling party on its toes with respect to electorate opinion.

35. There is little electoral harmonization between states or between state and national governments in Australia. Tasmania is an exception in terms of Green Party representation in the state legislature because it is the only state to run its elections on a completely proportional system.

36. Doyle and Kellow argue that this preferencing of prodevelopment policies can be ascribed to state governments' lack of access to the main macroeconomic instruments governments can use to their electoral advantage. Hence fostering such projects provides a way for them to deliver the goods, in terms of material benefits, to the voters (Doyle and Kellow 1995:129).

37. The Australian environmental movement, as is the case with many of its industrialized counterparts, is composed of a range of different groups, including wealthy, professional groups such as the Australian Conservation Foundation, as well as more direct action-oriented grassroots and community groups and a small Green Party. One noted characteristic of the Australian movement is its tendency to focus more on "green" than "brown" environmental issues (i.e., issues of resource conservation and wildlife protection rather than pollution control issues).

38. See also Robinson 1994.

39. They also ascribe an apparent decline in political support for environmental groups to a "bifurcation" of preferences: while the general public is more concerned about "brown" than "green" environmental issues, environmental groups tend to focus on the latter (Pakulski, Tranter, and Crook 1998:246–249).

40. Thanks to Alastair Iles for information here.

41. "No Chemical Storage Proposal for Point Lillias," Ramsar Archives News Release, June 24, 1997. See www2.iucn.org/themes/ramsar/w.n.lillias_release2.htm.

42. Three new processes are used by five firms to destroy certain scheduled waste: a hydrogenation process called Ecologic, which converts hazardous substances to methane, most of which is then converted to hydrogen, in use in Western Australia; Base Catalyzed Dechlorination (BCD), used in Queensland and Victoria; and Plasma Arc technology, in Victoria. Bioremediation, which involves the use of microorganisms (such as white-rot fungi) to break down toxic compounds, remains in the testing stage. See Scheduled Wastes Fact Sheet No. 8, June 1997, at http://www.environment.gov.au/epg/pubs/fs_scheduled8.html, and Review Report No. 4, *Appropriate Technologies for the Treatment of Scheduled Wastes*, prepared by CMPS&F Environmental, November 1997, available on Environet Australia at http://www.environment.gov.au/epg/environet/swtt/swtt.html.

43. See also Anderson 1992a and 1992b.

44. From Annex VII Issue Paper, presented at Basel Convention Technical Group Meeting, April 1999, courtesy of Geoff Thompson, Australian Environment Agency.

45. See http://www.environment.gov.au/epg/hwa/prg_permit.html. Of course, figures for only one year should not be taken as accurate for every year.

46. Vallette and Spalding (1990) cite Japan as neither exporting nor importing wastes; however, they do note government plans for exporting 40 tonnes of waste to the Marshall Islands in 1990. Japan does export very small quantities of wastes to various parts of Southeast Asia, primarily for recycling purposes.

47. These cases—two cases of mercury poisoning in the cities of Minamata and Niigata, one of cadmium poisoning, and a widespread outbreak of asthma in the Yokkaichi region—came to light during the late-1950s and early-1960s and became known in the 1970s as the "Big Four" lawsuits (see discussion below).

48. The first waste disposal and public cleansing law was enacted as part of the 1970 Pollution Diet; it has been subsequently amended in 1991 and 1992. Also in 1992 the Japanese introduced a law, which implemented the provisions of the Basel Convention, concerning the import and export of hazardous wastes. Concerning the definition of hazardous wastes, the OECD notes that "Japan classifies fewer items as hazardous than any other OECD country" (OECD 1993d:49). For example, only industrial wastes containing toxic compounds are considered "hazardous"; dioxins or PCBs in municipal wastes might be classed as "problem wastes" but not as "hazardous" (Naito 1987:161).

49. In his discussion of local-national government relations in Japan, Reed in fact categorizes the Japanese system as lying somewhere between the British-style unitary system and federal systems. The source of the relative autonomy of the prefectures lies in a constitutional grant of authority enabling prefectures to legislate independently of central government, but "within the limits of the law" (Reed 1986:24–26).

50. "Waste Management in Japan," *Haznews* 85, April 1995:18.

51. Although Delahunt and Burhenn note that the Japanese government also permits "coastal reclamation" as a means of disposing of many types of wastes, including industrial wastes. This technique involves the creation of additional land mass via the construction of waste-filled structures in the oceans just off the coasts; until 1976, these were not considered waste disposal facilities (Delahunt and Burhenn 1991:62). Japan also has significant problems with illegal waste dumping, arising at least in part from the severe constraints on landfill and incineration capacity—for example, 8,853 incidents of illegal dumping were reported in 1986 (Weidner 1989a:494). For a discussion of waste management at the municipal level, see Hershkowitz and Salerni 1987.

52. As witnessed in the failure of the Environment Agency to carry through legislation that would have introduced environmental impact assessment in Japan, a fight that began in 1981 (Crump 1996; Barrett and Therivel 1991).

53. Informal means include personal ties and the practice of *amakudari*—literally, "descent from heaven"—whereby top bureaucrats often retire to take up powerful positions in industry. See Boyd 1987.

54. Crump 1996 notes that although academics and representatives of environmental groups are sometimes brought in to testify before committees, their role is more often to "rubber stamp" draft legislation in sessions that are carefully scripted by the bureaucracy (Crump 1996:117).

55. Although five small parties with strong environmental representation in their platforms stood in the 1989 elections, garnering roughly 1.3% of all votes cast (Barrett and Therivel 1991:12). For a discussion of Japan's unique electoral system, see Stockwin 1982. In 1994 the electoral system was replaced by a mixed system, mirroring the German one, under which "voters ... cast two ballots, one for a political party in a national proportional representation system, and the other for a single candidate in a new one-member district system" (Blaker 1995:3).

56. This is in comparison with Germany, for example, where firms are held accountable to the courts and to public opinion through a thorough and effective system of monitoring and prosecution in the face of wrongdoing.

57. For a discussion of some of these, see Boyd 1987.

58. As witnessed by its ability to impose the technocratic policy measures discussed above, set environmental quality standards, and set up a sophisticated monitoring system (Weidner 1989a:492).

59. In fact the Japanese system has sometimes been characterized as "corporatism without labor" (Pempel and Tsunekawa 1979).

60. For an analysis of the early rise and subsequent decline of the Japanese environmental movement in comparison with the German Greens, see Schreurs 1997. For other works on the Japanese environmental movement, see McKean 1981, Tsuru and Weidner 1989, Cole 1994, Barrett and Therivel 1991, and Crump 1996.

61. Although this does not carry over to other international environmental issue areas, where Japan is usually considered to be a laggard (e.g., whaling, the wildlife trade) and has been accused by many of its neighbors of exporting environmental harm—via industrial relocation rather than waste exportation. For analyses of Japan's stance regarding international environmental issues and negotiations, see Schreurs 1997, Porter and Brown 1995, and Vogel 1992. For a critique of Japan's environmental foreign aid program, see Potter 1994, and for its industrial practices overseas, or its "shadow economy," see Dauvergne 1997.

62. For further discussion of the causes and social, economic, and health-related effects of Minamata disease and the other syndromes, see Tsuru 1989, Mishima 1992, and Ui 1992. It has been argued that one of the results of the regulations following these cases was that many of the main firms involved relocated some of their more dangerous activities to lower-profile countries of Southeast Asia (see Ui 1989).

258   Notes

63. At the same time, there are several disadvantages associated with this approach: citizen groups were often seen to have been "bought off" by compensation, and financial rewards have certainly contributed to the failure of the groups to coalesce into a more permanent and cross-sectoral movement. Also, Weidner emphasizes that the Japanese approach therefore focuses much more on ameliorating the effects of pollution than on reducing its generation.

64. In 1988 the compensation system was fundamentally revised to reduce the number and types of diseases covered and to try to develop a more comprehensive approach to pollution control and paying out compensation (Weidner 1989b:161).

65. For example, by fostering regional environmental cooperation in the Pacific area (see Potter 1994 and English and Runnalls 1997).

66. See Tsuru, Weidner 1989 and Barrett and Therivel 1991.

Chapter 7

1. These social costs include the actual or perceived environmental harm imposed on local communities and ecosystems and the longer term economic costs of cleaning up contaminated and abandoned disposal sites.

2. This observation holds across other OECD countries as well. See Yakowitz 1993.

3. See also Weale et al. 1996 on national environment ministries.

4. This corresponds with Young's notion of procedural effectiveness (Young 1994).

5. These issues are "highly" comparable in the sense that it would be possible to employ an extremely similar methodology in explaining them through identifying the main stakeholders and their preferences, then tracing out how the stakeholders go about achieving these preferences in the context of regulatory restraints and/or opportunities. See Clapp 1997 on the illegal CFC trade.

6. Some related works study the diffusion of ideas and the notion of social learning, and how governments and organizations themselves learn. Using the categories developed by Jänicke (1982), diffusion may happen "horizontally" (e.g., Sand 1990), or "vertically" through the network of international environmental regimes and agreements (e.g., Haas, 1999). Some of the work on advocacy coalitions and policy learning at the domestic level includes Sabatier and Jenkins-Smith 1993 and Lee 1993; thanks to Alastair Iles for conversations here. Lateral links between private actors across national boundaries are an important part of these processes, for example, between firms (as in ISO agreements), between other societal actors (transnational networks), and even between groups of local authorities in different countries.

7. See Vogel 1995:13–18 for a discussion of nontariff barriers to trade. The most important case in the EC/EU context is the Cassis de Dijon case (1979),

where the European Court of Justice ruled that the German minimum alcohol content ruling constituted an unfair barrier to trade. This ruling "made explicit the concept of 'mutual recognition'" of domestic differences (Vogel 1995:31). Although this seemed to allow for a less complex solution than harmonizing domestic health and safety standards, Vogel argues that in fact this judgment opened up the doors for such processes, leaving both consumers and many producers unwillingly exposed to "products produced according to the standards of the least stringent national authority" (ibid.:33).

8. Directive 96/61.

9. The European Environment Agency (established in 1994) spent its first years of existence collecting harmonized environmental data from the member states, a next to impossible task given the huge differences between national monitoring systems ("European Green Police Have Carrot but No Stick," *New York Times*, September 8, 1996:3). Its function is primarily as a monitoring agency; it has no formal authority to implement or enforce European environmental legislation.

10. Although for a more nuanced account of policy convergence and feedback —one that leaves room for national differences—see Coleman and Grant 1998.

11. Interestingly, although Britain ranks high on the list of states that infringe or fail to implement environmental directives, its record is by no means the worst overall, and in many cases Germany outranks it in this area. For example, one study that compares Germany's and Britain's record on environmental directive implementation shows a 20% deficit on the part of the Germans, compared to a 15.38% deficit on the part of the British (Peattie and Ringler 1994:218).

12. For example, in 1991 the Netherlands introduced a National Environmental Policy Plan—the most ambitious attempt yet by any country to introduce concepts of sustainability throughout the public policy sphere. See Wintle and Reeve 1994.

13. For overviews and critiques of ecological modernization theory, see Blowers 1997, Mol 1996, and Christoff 1996. The relationship between environment and modernity in the broadest senses arose from the work of Giddens (e.g., 1990) and Beck (1992, 1996). Christoff (1996) distinguishes between three uses of the term—as environmentally sensitive technological change, as a style of policy discourse, and as a belief system possibly connoting systemic change—in order to draw up a typology of weak versus strong ecological modernization. Work by Weale and Hajer (e.g., 1996) fits best into the policy discourse and belief system categories. Christoff's major critique of the concept, which is echoed by Blowers, is that "[it] may serve to legitimize the continuing instrumental domination and destruction of the environment, and the promotion of less democratic forms of government.... Consequently there is a need to identify the normative dimensions of these uses as either weak or strong, depending on whether or not such ecological modernization is part of the problem or part of the solution for the ecological crisis" (Christoff 1996:497).

14. See Weale 1992a:75–79, Christoff 1996:477.

15. Thus the ecological modernization framework shares much in common with other ideational approaches in vogue in the New Institutionalist field at the moment (e.g., Goldstein and Keohane 1993; Sikkink 1991; Hall 1992), as well as some of the same problems (Blyth 1997).

16. See Strohm 1993 and Copeland 1991.

17. "Cambodia Town's Luck Leaves Illness in its Wake," *New York Times*, January 4, 1999.

18. "Africa Study Reveals State of Hazwaste Management," *Haznews* 109, April 1997:17; other reports from "Chemcontrol Symposium II: Developing Markets," *Haznews* 104, November 1996:22–4.

19. "Proposed Landfill Directive Issued," *Haznews* 109, April 1997:1.

20. For example, the British National Association of Waste Disposal Contractors (NAWDC) is now renamed the Environmental Services Association. The environmental services industry is one of the biggest growth sectors in the world. In 1996 the OECD estimated the value of the global market at $250 billion (OECD 1996), predicting "that for waste management products, demand will grow by over 50 percent between 1990 and 2000, with the value of the world solid waste handling market estimated at $213bn by the year 2000" (Cooke and Chapple 1996:15). There is as yet no universally accepted definition of this industry. However, it is usually considered to include any goods or services, employed at all stages of the production process, considered better for the environment than their alternatives (OECD 1996:213–214).

21. See Delmas 1999 and Clapp 1997.

22. Agenda 21, on sustainable development, introduced at the 1992 UNCED meeting in Rio, represents the main attempt by the UNEP, the UNDP, international environmental NGOs and some governments to bring environmental and economic concerns under one common framework. Although it has received a high level of rhetorical support, it has yet to acquire this level of support, or even a shared understanding of its aims, in practice.

23. See Krueger 1998 for the argument that definitions under Basel should focus on "hazard" rather than "waste" to develop a coherent regulatory scheme.

24. See www.epa.gov/earth1r6/6en/h/haztraks/haztraks.htm. (January 1999)

25. The UNEP also sponsors the development of regional and subregional centers for training and technological transfers of practices regarding hazardous waste management and minimization.

26. See Forester and Skinner 1987, Batstone, Smith, and Wilson 1989, and Bromm 1990.

References

Aarhus, Knut. "Rhetoric and Reality in European Pollution Control." *International Environmental Affairs* 7.2 (1995): 111–126.

Abromeit, Heidrun. "Government-Industry Relations in West Germany." In *Governments, Industries, and Markets: Aspects of Government-Industry Relations in the U.K., Japan, West Germany, and the U.S.A. since 1945*. Ed. Martin Chick. Aldershot: Edward Elgar, 1990.

Adeola, Francis O. "Cross National Environmentalism Differentials: Empirical Evidence from Core and Noncore Nations." *Society and Natural Resources* 11.4 (1998): 339–364.

Aguilar, Susan. "Corporatist and Statist Designs in Environmental Policy: The Contrasting Roles of Germany and Spain in the EC Scenario." *Environmental Politics* 2.2 (1993): 223–47.

Allen, Robert. *Waste Not, Want Not: The Production and Dumping of Toxic Waste*. London: Earthscan, 1992.

Alter, Harvey. "Industrial Recycling and the Basel Convention." *Resources, Conservation, and Recycling* 19.1 (1997): 29–53.

Anderson, Ian. "Toxic Waste Exports 'Morally Justified'." *New Scientist* 135.1840 (1992a): 9.

Anderson, Ian. "While Australia Questions its Trade in Toxic Waste." *New Scientist* 136.1846 (1992b): 8.

Anton, Donald K., Jennifer Kohout, and Nicola Pain. "Nationalizing Environmental Protection in Australia: The International Dimensions." *Environmental Law* 23 (1993): 763–83.

Armour, Audrey M. "Risk Assessment in Environmental Policymaking." *Policy Studies Review* 12.3/4 (1993): 178–96.

Asante-Duah, D. Kofi, F. Frank Saccomano, and John H. Shortreed. "The Hazardous Waste Trade: Can it be Controlled?" *Environmental Science and Technology* 26.9 (1992): 1684–93.

Atkinson, Michael M., and William D. Coleman. "Strong States and Weak States: Sectoral Policy Networks in Advanced Industrialized Countries." *British Journal of Political Science* 19 (1989): 47–67.

Auer, Matthew R. "Negotiating Toxic Risks: A Case from the Nordic Countries." *Environmental Politics* 5.4 (1996): 687–699.

Baldwin, Paul. "Innovative Technology, Competitiveness, and Policy Choices at International Environmental Negotiations." Ph.D. dissertation. Columbia University, 1999.

Barnes, P. M. L. "Hazardous Waste Disposal in the U.K." *International Journal of Environment and Pollution* 7.2 (1997): 169–204.

Barrett, Brendan F. D., and Riki Therivel. *Environmental Policy and Impact Assessment in Japan*. London: Routledge, 1991.

Barta, C. "Hazardous Waste Disposal in Germany: A Case Study of the Site in Billigheim, Baden Württemberg." *International Journal of Environment and Pollution* 7.2 (1997): 154–168.

Batstone, Roger, James E. Smith, Jr., and David Wilson, eds. *The Safe Disposal of Hazardous Wastes: The Special Needs and Problems of Developing Countries*. 3 vols. Washington, DC: World Bank, 1989.

Beck, Ulrich. *Ecological Politics in an Age of Risk*. Trans. Amos Weisz. Cambridge: Polity Press, 1995.

Beck, Ulrich. "Risk Society and the Provident State." In *Risk, Environment and Modernity: Towards a New Ecology*. Eds. Scott Lash, Bronislaw Szersynski, and Brian Wynne. London: Sage Publications, 1996.

Beck, Ulrich. *Risk Society: Towards a New Modernity*. London: Sage, 1992.

Beder, Sharon. "The Burning Issue of Australia's Toxic Waste." *New Scientist* 130.1772 (1991): 35–38.

Beecher, Norman, and Ann Rappaport. "Hazardous Waste Management Policies Overseas." *Chemical Engineering Progress* 86 (1990): 30–39.

Benedick, Richard. *Ozone Diplomacy*. Cambridge, MA: Harvard University Press, 1987.

Bernauer, Thomas. "The Effect of International Environmental Institutions: How We Might Learn More." *International Organization* 49.2 (1995): 351–77.

Biod, Alexandra, Jane Probert, and Clifford Jones. "EU Waste Management: Lessons to be Learnt from French Experience?" *European Environment* 4.3 (1994): 21–25.

Birnie, Patricia W., and Alan E. Boyle. *International Law and the Environment*. Oxford: Oxford University Press, 1992.

Blaker, Michael. "Japan in 1994: Out with the Old, In with the New?" *Asian Survey* 35, 1 (1995): 1–12.

Block, Alan A. *Space, Time, and Organized Crime*. New Brunswick: Transaction Publishers, 1994.

Blowers, Andrew. "Environmental Policy: Ecological Modernization or the Risk Society?" *Urban Studies* 34.5/6 (1997): 845–871.

Blowers, Andrew. "Transboundary Transfers of Hazardous and Radioactive Wastes." In *Environmental Problems as Conflicts of Interest*. Eds. Peter Sloep and Andrew Blowers. London: Arnold, 1996.

Blowers, Andrew. "Transition or Transformation: Environmental Policy Under Thatcher." *Policy Administration* 65 (1987): 241–77.

Blowers, Andrew, David Lowry, and Barry D. Solomon. *The International Politics of Nuclear Waste*. London: MacMillan, 1991.

Blühdorn, Ingolfur. "Campaigning for Nature: Environmental Pressure Groups in Germany and Generational Change in the Ecology Movement." In *The Green Agenda: Environmental Politics and Policy in Germany*. Eds. Ingolfur Blühdorn, Frank Krause, and Thomas Scharf. Keele: Keele University Press, 1995.

Blühdorn, Ingolfur, Frank Krause, and Thomas Scharf, eds. *The Green Agenda: Environmental Politics and Policy in Germany*. Keele: Keele University Press, 1995.

Blyth, Mark M. "Any More Bright Ideas? The Ideational Turn of Comparative Political Economy." *Comparative Politics* 29.2 (1997): 229–250.

Bodiguel, Maryvonne, and Henry Buller. "Environmental Policy and the Regions in France." *Regional Politics and Policy* 4.3 (1994): 92–109.

Boehmer-Christiansen, Sonja, and Jim Skea. *Acid Politics: Environmental Politics and Energy Policies in Britain and Germany*. London: Belhaven Press, 1991.

Boehmer-Christiansen, Sonja, and Helmut Weidner. *The Politics of Reducing Vehicle Emissions in Britain and Germany*. London: Pinter, 1995.

Bogdanor, Vernon. *Democracy and Elections: Electoral Systems and Their Political Consequences*. Cambridge: Cambridge University Press, 1983.

Bothe, Michael. *Verwaltungsorganisation im Umweltschutz: Ressortzuständigkeiten und Sonderbehörden*. Linz: Institut für Kommunalwissenschaften und Umweltschutz, 1986.

Bowman, A. O. "Hazardous Waste Management: An Emerging Area within an Emerging Federalism." *Publius: The Journal of Federalism* 15 (1985): 131–144.

Boyd, Richard. "Government-Industry Relations in Japan: Access, Communication, and Competitive Collaboration." In *Comparative Government-Industry Relations: Western Europe, the United States and Japan*. Eds. Stephen Wilks and Maurice Wright. Oxford: Clarendon Press, 1987.

Braden, John B., Hank Folmer, and Thomas S. Ulen, eds. *Environmental Policy with Political and Economic Integration: The European Union and the United States*. Cheltenham: Edward Elgar, 1996.

Brechin, Steven R., and Willett Kempton. "Global Environmentalism: A Challenge to the Postmaterialism Thesis?" *Social Science Quarterly* June 1994: 245–269.

Brickman, Ronald, Sheila Jasanoff, and Thomas Ilgen. *Controlling Chemicals: The Politics of Regulation in Europe and the United States*. Ithaca: Cornell University Press, 1985.

Bromm, Susan. "Creating a Hazardous Waste Management Program in a Developing Country." *American Journal of International Law and Policy* 5 (1990): 325–50.

Brusco, Sebastiano, Paolo Bertossi, and Alberto Cottica. "Playing on Two Chessboards—The European Waste Management Industry: Strategic Behavior in the Market and the Policy Debate." In *Environmental Policy in Europe*. Ed. Francois Lévêque. Cheltenham: Edward Elgar, 1996.

Buell, Lawrence. "Toxic Discourse." *Critical Inquiry* 24.3 (1998): 639–665.

Bullard, Robert. *Dumping in Dixie: Race, Class, and Environmental Quality*. Boulder, CO: Westview Press, 1991.

Buller, Henry. "Reflections across the Channel: Britain, France, and the Europeanization of National Environmental Policy." In *British Environmental Policy and Europe: Politics and Policy in Transition*. Eds. Philip Lowe and Stephen Ward. London: Routledge, 1998, pp. 67–84.

Buller, Henry, Philip Lowe, and Andrew Flynn. "National Responses to Europeanization of Environmental Policy: A Selective Review of Comparative Research." In *European Integration and Environmental Policy*. Eds. J. D. Lieferink, P. D. Lowe and A. P. J. Mol. London: Belhaven Press, 1993, pp. 175–195.

Bulmer, Simon. "Territorial Government." In *Developments in West German Politics*. Eds. Gordon Smith, William E. Paterson, and Peter H. Merkl. London: MacMillan, 1989.

Cammack, Paul. "The New Institutionalism: Predatory Rule, Institutional Persistence, and Macro-Social Change." *Economy and Society* 21.4 (1992): 397–429.

Caporaso, James A., and Joseph Jupille. "The Europeanization of Social Policy and Domestic Political Change." Presented at the Europeanization and Domestic Political Change—3rd Project Workshop, European University Institute, Florence, 1998.

Carson, Rachel. *Silent Spring*. Cambridge, MA: Riverside Press, 1962.

Carter, Neil, and Philip Lowe. "Environmental Politics and Administrative Reform." *Political Quarterly* 65.3 (1994): 263–274.

Cash, David, and Susanne Moser. "Information and Decision Making Systems for the Effective Management of Cross-Scale Environmental Problems: A Theoretical Concept Paper." Global Environmental Assessment Project Working Paper, BCSIA, Kennedy School of Government, Harvard University, 1998.

Chapman, Graham, Kevel Kumar, Caroline Fraser, and Ivor Gaber. *Environmentalism and the Mass Media: The North-South Divide*. London: Routledge, 1997.

Chayes, Abram, and Antonia H. Chayes. "Compliance without Enforcement: State Behaviour under Regulatory Treaties." *Negotiation Journal* 7 (1991): 311–330.

Chayes, Abram, and Antonia H. Chayes. "On Compliance." *International Organization* 47.2 (1993): 175–205.

Christoff, Peter. "Ecological Modernization, Ecological Modernities." *Environmental Politics* 5.3 (1996): 476–500.

Christoff, Peter. "Environmental Politics." In *Developments in Australian Politics*. Eds. Judith Brett, James Gillespie, and Murray Groot. Melbourne: Macmillan Education Australia, 1994.

Clapp, B. W. *An Environmental History of Britain since the Industrial Revolution*. London: Longman, 1994.

Clapp, Jennifer. "Foreign Direct Investment in Hazardous Industries in Developing Countries." *Environmental Politics* 7.4 (1998): 92–113.

Clapp, Jennifer. "Global Industry Environmental Standards: Green Saviour for the South?" Presented at the Annual Meeting of the International Studies Association, Toronto, 1997.

Clapp, Jennifer. "The Global Recycling Industry and Hazardous Waste Trade Facilities." Presented at the Annual Meeting of the International Studies Association, Washington, DC, 1999.

Clapp, Jennifer. "The Illegal CFC Trade: An Unexpected Wrinkle in the Ozone Protection Regime." *International Environmental Affairs* 9, 4 (1997): 259–273.

Cole, Alistair, and Brian Doherty. "France: Pas commes les autres—the French Greens at the Crossroads." In *The Green Challenge: The Development of Green Parties in Europe*. Eds. Dick Richardson and Chris Rootes. London: Routledge, 1995.

Cole, Jonathan A. "The Right to Demand: Citizen Activism and Environmental Politics in Japan." *Journal of Environment and Development* 3.2 (1994): 77–95.

Coleman, W. D., and W. P. Grant. "Policy Convergence and Policy Feedback: Agricultural Finance Policies in a Globalizing Era." *European Journal of Political Research* 34.2 (1998): 225–247.

Conca, Ken, and Ronnie D. Lipschutz. "A Tale of Two Forests." In *The State and Social Power in Global Environmental Politics*. Eds. Ronnie D. Lipschutz and Ken Conca. New York: Columbia University Press, 1993.

Cooke, Andrew, and Wendy Chapple. "EU Regulation and the U.K. Waste Disposal Industry." *Environmental Policy and Practice* 6.1 (1996): 11–16.

Copeland, Brian. "International Trade in Waste Products in the Presence of Illegal Disposal." *Journal of Environmental Economics and Management* 20 (1991): 143–162.

Cothern, C. Richard. *Handbook for Environmental Risk Decision Making: Values, Perceptions, and Ethics*. Boca Raton: Lewis Publishers, 1996.

Crooks, Harold. *Giants of Garbage: The Rise of the Global Waste Industry and the Politics of Pollution Control*. Toronto: James Lorimer and Company, 1993.

Crump, John. "Environmental Politics in Japan." *Environmental Politics* 5.1 (1996): 115–121.

Cubel-Sánchez, Pablo. "Transboundary Movements of Hazardous Wastes in International Law: The Special Case of the Mediterranean Area." *International Journal of Coastal and Marine Law* 12.4 (1997), pp. 447–487.

Czech, Herbert. "Classification and Underground Disposal of Hazardous Wastes in Germany." Presented at the GTDC Workshop, Vienna, Austria, 1996.

Dalal-Clayton, Barry. *Getting to Grips with Green Plans: National-Level Experience in Industrial Countries*. London: Earthscan, 1996.

Dalton, Russell J., and Manfred Kuechler. *Challenging the Political Order: New Social and Political Movements in Western Democracies*. New York: Oxford University Press, 1990.

Dauvergne, Peter. *Shadows in the Forest: Japan and the Politics of Timber in Southeast Asia*. Cambridge, MA: MIT Press, 1997.

Davis, Gary, Donald Huisingh, and Bruce Piasecki. "Waste Reduction Strategies: European Practice and American Prospects." In *America's Future in Toxic Waste Management: Lessons from Europe*. Eds. Bruce W. Piasecki and Gary Davis. New York: Quorum Books, 1987.

Defregger, Franz. "Status and Trends on the Management of Industrial Hazardous Waste in the Federal Republic of Germany." *UNEP Industry and Environment* Special Issue (1983): 15–21.

Delahunt, Katherine W., and David Burhenn. "Japan: Environmental Risk." *International Financial Law Review (Special Supplement)* February (1991): 57–63.

Delmas, Magali A. "In Search of ISO: Barriers and Incentives to the Globalization of Environmental Management Standards. The Case of the United States." (1999), paper presented at Workshop on Globalization and Regulation Centers for German and European Studies and Law and Society, UC Berkeley, May 1999.

Deloria, Vine Jr., and Clifford M. Lytle. *American Indians, American Justice*. Austin: University of Texas Press, 1983.

Department of the Environment, *Waste Management, Planning: Principles and Practice* (London: HMSO, 1995).

Department of the Environment, *United Kingdom Management Plan for Exports and Imports of Waste* (London: HMSO, 1996).

Desai, Uday, ed. *Ecological Policy and Politics in Developing Countries: Economic Growth, Democracy, and Environment*. Albany: SUNY Press, 1998.

Dicken, P. "International Production in a Volatile Regulatory Environment: the Influence of National Regulatory Structures on the Spatial Strategies of Transnational Corporations." *Geoforum* 23.3 (1992): 303–316.

Dobson, Andrew. *Green Political Thought*. 2nd ed. London: Routledge, 1995.

Douence, Jean-Claude. "The Evolution of the 1982 Regional Reforms: An Overview." *Regional Politics and Policy* 4.3 (1994): 10–24.

Dowling, Michael, and Joanne Linnerooth. "The Listing and Classifying of Hazardous Wastes." In *Risk Management and Hazardous Waste: Implementation and the Dialectics of Credibility*. Ed. Brian Wynne. Berlin: Springer-Verlag, 1987.

Downes, David. "Neo-Corporatism and Environmental Policy." *Australian Journal of Political Science* 31.2 (1996): 175–190.

Downie, David. "Road Map or Dead-End: Is the Montreal Protocol a Precedent for Global Climate Change Negotiations?" Presented at the Annual Meeting of the American Political Science Association, New York, 1994.

Doyle, Timothy, and Aynsley Kellow. *Environmental Politics and Policy-Making in Australia*. Melbourne: MacMillan, 1995.

Dryzek, John. *Rational Ecology: Environment and Political Economy*. Oxford: Basil Blackwell, 1987.

Dunlap, Riley E., George H. Gallup, Jr., and Alec M. Gallup. "Of Global Concern: Results of the Health of the Planet Survey." *Environment* November (1993): 7–15, 33–39.

Dunlap, Riley E., and Angela G. Mertig. "Global Environmental Concern: An Anomaly for Postmaterialism." *Social Science Quarterly* (March 1997): 24–29.

Dyson, Ken, ed. *The Politics of German Regulation*. Aldershot: Dartmouth Publishing Company, 1992.

Easterbrook, Gregg. "Forget PCB's, Radon, Alar." *New York Times Magazine* September 11, 1994, pp. 60–63.

Eckersley, Robyn. *Environmentalism and Political Theory: Towards an Ecocentric Approach*. Albany: SUNY Press, 1992.

Eckersley, Robyn, ed. *Markets, the State and the Environment: Towards Integration*. London: MacMillan Press Ltd., 1996.

Elman, Miriam Fendius. "The Foreign Policies of Small States: Challenging Neorealism in its own Backyard." *British Journal of Political Science* 25 (1995): 171–217.

English, H. Howard, and David Runnalls, eds. *Environment and Development in the Pacific: Problems and Policy Options*. Melbourne: Addison-Wesley, 1997.

Enloe, Cynthia. *The Politics of Pollution in a Comparative Perspective: Ecology and Power in Four Nations*. New York: David MacKay Company, 1975.

Entwistle, Tom. "Trust in Waste Regulation?" *Environmental Politics* 6.4 (1997): 179–187.

Environmental Data Service, London, *ENDS Report*, various issues, 1985–1999.

Environmental Protection Agency. *Unfinished Business: A Comparative Assessment of Environmental Problems*. Washington, DC: Environmental Protection Agency, 1987.

Faber, Daniel, ed. *The Struggle for Ecological Democracy: Environmental Justice Movements in the United States*. New York: The Guilford Press, 1998.

Farrell, Alex, and Terry J. Keating. "Multi-Jurisdictional Air Pollution Assessment: A Comparison of the Eastern United States and Western Europe." ENRP Discussion Paper E-98-12, Kennedy School of Government, Harvard University, 1998.

Faucher, Florence, and Brian Doherty. "The Decline of Green Politics in France: Political Ecology since 1992." *Environmental Politics* 5.1 (1996): 108–114.

Finkel, Adam M., and Dominic Golding, eds. *Worst Things First? The Debate over Risk-Based National Environmental Priorities*. Washington, DC: Resources for the Future, 1994.

Flynn, Andrew, and Philip Lowe. "The Greening of the Tories: The Conservative Party and the Environment." In *Green Politics Two*. Ed. Wolfgang Rüdig. Edinburgh: Edinburgh University Press, 1992.

Forester, William S., and John H. Skinner, eds. *International Perspectives on Hazardous Waste Management*. London: Academic Press, 1987.

Fox, Jonathan A., and L. David Brown, eds. *The Struggle for Accountability: The World Bank, NGOs, and Grassroots Movements*. Cambridge, MA: MIT Press, 1998.

Frankland, E. Gene. "Germany: The Rise, Fall, and Recovery of Die Grünen." In *The Green Challenge: The Development of Green Parties in Europe*. Eds. Dick Richardson and Chris Rootes. London: Routledge, 1995.

Garner, Robert. *Environmental Politics*. London: Prentice Hall/Harvester Wheatsheaf, 1996.

Giddens, Anthony. *The Consequences of Modernity*. Oxford: Polity Press, 1990.

Gilpin, Robert. *The Political Economy of International Relations*. Princeton: Princeton University Press, 1987.

Goldstein, Judith, and Robert O. Keohane, eds. *Ideas and Foreign Policy: Beliefs, Institutions, and Political Change*. Ithaca: Cornell University Press, 1993.

Golub, Jonathan. "EC Environmental Policy before and after the Single European Act: The Pivotal Role of British Sovereignty." Presented at Shape of the New Europe: The Eleventh Annual Graduate Student Conference, Institute on Western Europe, Columbia University, New York, 1994.

Gordon, John. "Environmental Policy in Britain and Germany: Some Comparisons." *European Environment* 4.3 (1994): 9–12.

Gourlay, K. A. *World of Waste: Dilemmas of Industrial Development*. London: Zed Books, 1992.

Grant, Wyn. *Pressure Groups, Politics, and Democracy in Britain*. New York: Harvester Wheatsheaf, 1995.

Grant, Wyn, and William Paterson. "The Chemical Industry: A Study in Internationalization." In *Governing Capitalist Economies: Performance and Control* of Economic Sectors. Eds. R. Hollingsworth, P. Schmitter, and W. Streeck. New York: Oxford University Press, 1994.

Haas, Peter M. "Obtaining Environmental Protection through Epistemic Consensus." Millennium 19.3 (1990a): 347–363.

Haas, Peter M. Saving the Mediterranean: The Politics of International Environmental Cooperation. New York: Columbia University Press, 1990b.

Haas, Peter M. "Social Constructivism and the Evolution of Multilateral Environmental Governance." In Globalization and Governance. Eds. Aseem Prakash and Geoffrey A. Hart. London: Routledge, 1999.

Haas, Peter M., Robert O. Keohane, and Marc A. Levy, eds. Institutions for the Earth: Sources of Effective International Environmental Protection. Cambridge, MA: MIT Press, 1993.

Hackett, David P. "An Assessment of the Basel Convention on the Control of Transboundary Movement of Hazardous Wastes and Their Disposal." American University Journal of International Law and Policy 5 (1990): 291–323.

Hager, Carol J. Technological Democracy: Bureaucracy and Citizenry in the German Energy Debate. Ann Arbor: University of Michigan Press, 1995.

Haggard, Stephan. Pathways from the Periphery: The Politics of Growth in the Newly Industrializing Countries. Ithaca: Cornell University Press, 1990.

Hahn, Robert W., and Kenneth R. Richards. "The Internationalization of Environmental Regulation." Harvard International Law Journal 30.2 (1989): 421–46.

Haigh, Nigel, and Frances Irwin, eds. Integrated Pollution Control in Britain and North America. Washington, DC: The Conservation Foundation, 1990.

Haigh, Nigel, and Chris Lanigan. "Impact of the European Union on UK Environmental Policy Making." In UK Environmental Policy in the 1990s. Ed. Tim S. Gray. London: MacMillan Press, 1995.

Hajer, Maarten A. "Ecological Modernization as Cultural Politics." Risk, Environment, and Modernity: Towards a New Ecology. Eds. Scott Lash, Bronislaw Szersynski, and Brian Wynne. London: Sage Publications, 1996.

Hall, Peter. "From Keynesianism to Monetarism: Institutional Analysis and British Economic Policy in the 1970s." In Structuring Politics: Historical Institutionalism in Comparative Perspective. Eds. Sven Steinmo, Kathleen Thelen, and Frank Longstreth. Cambridge: Cambridge University Press, 1992.

Hall, Peter. Governing the Economy: The Politics of State Intervention in Britain and France. New York: Oxford University Press, 1986.

Hall, Peter A. "The Role of Interests, Institutions, and Ideas in the Comparative Political Economy of the Industrialized Nations." In Comparative Politics: Rationality, Culture and Structure. Eds. Mark Irving Lichbach and Alan S. Zuckerman. Cambridge: Cambridge University Press, 1997.

Hall, Peter A., Jack Hayward, and Howard Machin, eds. Developments in French Politics. Rev. ed. London: MacMillan, 1994.

Handl, Günther. "Environmental Protection and Development in Third World Countries: Common Destiny—Common Responsibility." *New York University Journal of International Law and Public Policy* 20.3 (1988): 603–627.

Handley, F. J. "Hazardous Waste Exports: A Leak in the System of International Legal Controls." *Environmental Law Reporter* 19, 4 (1989): 10171–10182.

Hanf, Kenneth, and Theo A. J. Toonen, eds. *Policy Implementation in Federal and Unitary Systems: Questions of Analysis and Design*. Dordrecht: Martinus Nijhoff Publishers, 1985.

Harr, Jonathan. *A Civil Action*. New York: Random House, 1995.

Haznews (London), various issues.

Hazardous Wastes Inspectorate, First Report: *Hazardous Waste Management—An Overview*, June 1985 (London: HMSO).

Heidenheimer, Arthur J., Hugh Heclo, and Carolyn Teich Adams, eds. *Comparative Public Policy: The Politics of Social Choice in America, Europe, and Japan*. 3rd ed. New York: St. Martin's Press, 1990.

Heller, Paul. *Database of Known Hazardous Waste Exports from OECD to Non-OECD Countries, 1989–March 1994*. Washington, DC: Greenpeace, 1994.

Helm, Dieter, and David Pearce. "Economic Policy Towards the Environment." *Oxford Review of Economic Policy* 7.4 (1990): 1–16.

Héritier, Adrienne. "The Accommodation of Diversity in European Policy Making and its Outcomes: Regulatory Policy as a Patchwork." *Journal of European Public Policy* 3.2 (1996): 149–167.

Héritier, Adrienne. "Differential Europe: National Administrative Responses to Community Policy." Presented at the Europeanization and Domestic Political Change—3rd Project Workshop, European University Institute, Florence, 1998.

Héritier, Adrienne, Suzanne Mingers, Christoph Knill, and Martina Becka. *Die Veränderung von Staatlichkeit in Europa—Ein regulativer Wettbewerb: Deutschland, Gross-Britannien und Frankreich in der Europaischen Union*. Opladen: Leske + Budrich, 1994.

Hershkowitz, Allen, and Eugene Salerni. *Garbage Management in Japan: Leading the Way*. New York: INFORM, 1987.

Herzik, Eric B., and Alvin H. Mushkatel, eds. *Problems and Prospects for Nuclear Waste Disposal Policy*. Westport, CT: Greenwood Press, 1993.

Hertzman, Clyde, and Aleck Ostry. "Community Risk Perception and Waste Management in Three Communities at Different Stages in the Siting Process." In *Hazardous Waste Siting and Democratic Choice*. Ed. Don Munton. Washington, DC: Georgetown University Press, 1996.

Heuter, Horst. "Full-Employment, Social Justice and, Sustainable Development: The Environmental Policy of the German Trade Union Association (DGB)." In *The Green Agenda: Environmental Politics and Policy in Germany*. Eds. Ingolfur Blühdorn, Frank Krause, and Thomas Scharf. Keele: Keele University Press, 1995.

Hildebrandt, Eckart. "Preconditions and Possibilities for a Trade Union Environmental Policy." *The Green Agenda: Environmental Politics and Policy in Germany.* Eds. Ingolfur Blühdorn, Frank Krause, and Thomas Scharf. Keele: Keele University Press, 1995.

Hildyard, Nicholas. "Down in the Dumps: Britain and Hazardous Wastes." In *Green Britain or Industrial Wasteland?* Eds. Edward Goldsmith and Nicholas Hildyard. London: Polity Press, 1986.

Hilz, Christoph. *The International Toxic Waste Trade.* New York: Van Nostrand Reinhold, 1992.

Hilz, Christoph, and J. R. Ehrenfeld. "Transboundary Movements of Hazardous Wastes: A Comparative Analysis of Policy Options to Control the International Waste Trade." *International Environmental Affairs* 3.1 (1991): 26–63.

Hilz, Christoph, and Mark Radka. "The Basel Convention on Transboundary Movement of Hazardous Waste and their Disposal." In *Nine Case Studies in International Environmental Negotiation.* Eds. L. Susskind, E. Siskind, and J. W. Breslin. Cambridge, MA: MIT-Harvard Public Disputes Program, 1990.

Holland, Kenneth M., F. L. Morton, and Brian Galigan, eds. *Federalism and the Environment: Environmental Policymaking in Australia, Canada and the United States.* Westport: Greenwood Press, 1996.

Hollingsworth, J. Rogers, Philippe C. Schmitter, and Wolfgang Streeck, eds. *Governing Capitalist Economies: Performance and Control of Economic Sectors.* New York: Oxford University Press, 1994.

House of Lords Select Committee on Science and Technology, chaired by Lord Gregson (1981). *Hazardous Waste Disposal,* 3 vols. (HMSO London).

Hunt, M. C., and J. A. Chandler. "France." In *Local Governments in Liberal Democracies: An Introductory Survey.* Ed. J. A. Chandler. London: Routledge, 1993.

Hurrell, Andrew, and Benedict Kingsbury, eds. *The International Politics of the Environment.* Oxford: Oxford University Press, 1992.

Hutter, Bridget M., ed. *A Reader in Environmental Law.* Oxford: Oxford University Press, 1999.

Hutter, Bridget M. *The Reasonable Arm of the Law? The Law Enforcement Procedures of Environmental Health Officers.* Oxford: Clarendon Press, 1988.

Hyman, Mark. "Australian Hazardous Waste Legislation and Our International Obligations." Presented at the Infrastructure 1997 Conference, Sydney, 1997.

Ikenberry, G. John. "Conclusion: An Institutional Approach to American Foreign Economic Policy." *International Organization* 42.1 (1988): 219–243.

Ikenberry, G. John, David A. Lake, and Michael Mastanduno, eds. *The State and American Foreign Economic Policy.* Ithaca: Cornell University Press, 1988.

Inglehart, Ronald. *Changing Values and the Rise of Environmentalism in Western Societies*. Berlin: International Institute for Environment and Society, 1982.

Irwin, Frances. "Introduction to Integrated Pollution Control." In *Integrated Pollution Control in Europe and North America*. Eds. Nigel Haigh and Frances Irwin. Washington, DC: The Conservation Foundation, 1990.

Jacobs, Scott H. "Regulatory Co-Operation for an Interdependent World: Issues for Government." In *Regulatory Co-Operation for an Interdependent World*. Paris: OECD, 1994.

Jaffe, Adam B., Steven R. Peterson, and Paul R. Portney. "Environmental Regulation and the Competitiveness of US Manufacturing: What Does the Evidence Tell Us?" *Journal of Economic Literature* 33 (1995): 132–163.

Jahn, Detlef. "Environmental Performance and Policy Regimes: Explaining Variations in 18 OECD Countries." *Policy Sciences* 31.2 (1998): 107–131.

Jänicke, Martin. "Conditions for Environmental Policy Success: An International Comparison." *The Environmentalist* 12.1 (1992): 47–58.

Jänicke, Martin, and Helmut Weidner, eds. *National Environmental Policies: A Comparative Study of Capacity Building*. Berlin: Springer-Verlag, 1997.

Jänicke, Martin, and Helmut Weidner, eds. *Successful Environmental Policy: A Critical Evaluation of 24 Cases*. Berlin: Rainer Bohn Verlag, 1995.

Jasanoff, Sheila. "Civilization and Madness: The Great BSE Scare of 1996." *Public Understanding of Science* 6.3 (1997): 221–232.

Jasanoff, Sheila. "Cross-National Differences in Policy Implementation." *Evaluation Review* 15.3 (1991): 103–119.

Jasanoff, Sheila. *Risk Management and Political Culture: A Comparative Study of Science in the Policy Context*. New York: Russell Sage Foundation, 1986.

Jordan, Andrew. "The Impact on U.K. Environmental Administration." In *British Environmental Policy and Europe: Politics and Policy in Transition*. Eds. Philip Lowe and Stephen Ward. London: Routledge, 1998a.

Jordan, Andrew. "EU Environmental Policy at 25: The Politics of Multinational Governance." *Environment* 40.1 (1998b): 14–20, 39–45.

Jordan, Andrew. "Integrated Pollution Control and the Evolving Style and Structure of Environmental Regulation in the U.K." *Environmental Politics* 2.3 (1993): 405–427.

Jordan, Andrew, and Timothy O'Riordan. "The Precautionary Principle in U.K. Environmental Policy." In *U.K. Environmental Policy in the 1990s*. Ed. Tim S. Gray. London: MacMillan, 1995.

Judge, David. *The Parliamentary State*. London: Sage, 1993.

Jupille, Joseph Henri. "Free Movement of Goods and Hazardous Waste: Reconciling the Single Market with Environmental Imperatives." Presented at the Annual Meeting of the International Studies Association, San Diego, 1996.

Kabala, S. "The Hazardous Waste Problem in Eastern Europe." *Report on Eastern Europe* June 21 (1991): 27–33.

Kagan, Robert A. "Trying to Have it Both Ways: Local Discretion, Central Control, and Adversarial Legalism in American Environmental Regulation." *Ecology Law Quarterly* 25 (1999): 718–732.

Kamieniecki, Sheldon, ed. *Environmental Politics in the International Arena: Movements, Parties, Organizations, and Policy.* Albany: SUNY Press, 1993.

Kasperson, Roger E., Dominic Golding, and Jeanne X. Kasperson. "Risk, Trust, and Democratic Theory." Presented at the Weatherhead Center for Science and International Affairs Seminar Series on International Environmental Politics, Harvard University, Cambridge, MA, 1998.

Katzenstein, Peter J. *Between Power and Plenty: Foreign Economic Policies of Advanced Industrial States.* Madison: University of Wisconsin Press, 1978.

Katzenstein, Peter J. *Small States in World Markets: Industrial Policy in Europe.* Ithaca: Cornell University Press, 1985.

Keck, Margaret E., and Kathryn Sikkink. *Activists Beyond Borders: Advocacy Networks in International Politics.* Ithaca: Cornell University Press, 1998.

Kelemen, R. Daniel. "Regulatory Federalism: EU Environmental Regulation in Comparative Perspective." Presented at the Annual Meeting of the American Political Science Association, Boston, MA, 1998.

Kellow, Aynsley. "Thinking Globally and Acting Federally: Intergovernmental Relations and Environmental Protection in Australia." In *Federalism and the Environment: Environmental Policymaking in Australia, Canada, and the United States.* Eds. Kenneth M. Holland, F. L. Morton and Brian Galligan. Westport: Greenwood Press, 1996

Kemp, R. *The Politics of Radioactive Waste Disposal.* Manchester: Manchester University Press, 1992.

Keohane, Robert O., and Marc A. Levy, eds. *Institutions for Environmental Aid: Pitfalls and Promise.* Cambridge, MA: MIT Press, 1996.

Kerr, Clark. *The Future of Industrial Societies: Convergence or Diversity?* Cambridge, MA: Harvard University Press, 1983.

Kitschelt, Herbert P. "Political Opportunity Structures and Political Protest: Anti-Nuclear Movements in Four Democracies." *British Journal of Political Science* 16 (1986): 57–85.

Knill, Christoph, and Andrea Lenschow. "Change as 'Appropriate Adaptation': Administrative Adjustment to European Environmental Policy in Britain and Germany." Presented at the Europeanization and Domestic Political Change—3rd Project Workshop, European University Institute, Florence, 1998.

Knoepfel, Peter, Lennart Lundqvist, Rémy Prudhomme, and Peter Wagner. "Comparing Environmental Policies: Different Styles, Similar Content." In *Comparative Policy Research: Learning from Experience.* Eds. Meinoff Dierkes, Hans N. Weiler, and Ariane Berthoin Antal. Berlin: WZB-Publications, 1987.

Knoepfel, Peter, and Helmut Weidner. "Explaining Differences in the Performance of Clean Air Policies: An International and Interregional Comparative Study." *Policy and Politics* 14.1 (1986): 71–91.

Koss, Klaus-Dieter, Ulrich Malorny, and Wilfried Stahlke. "Europäischer Abfallkatalog und die EG-Abfallverbringungs-Verordnung." *Müll und Abfall* 10 (1994): 625–640.

Krasner, Stephen D. *Defending the National Interest.* Princeton: Princeton University Press, 1978.

Krasner, Stephen D., ed. *International Regimes.* Ithaca: Cornell University Press, 1983.

Krasner, Stephen D. "Sovereignty: An Institutional Perspective." *Comparative Political Studies* 21.1 (1988): 66–94.

Krueger, Jonathan. *International Trade and the Basel Convention.* London: Royal Institute of International Affairs/Earthscan Press, 1999.

Krueger, Jonathan. "When Is a Waste Not a Waste? The Evolution of the Basel Convention and the International Trade in Hazardous Wastes." Presented at the Annual Meeting of the International Studies Association, Minneapolis, MN, 1998.

La Porte, Todd R., and Daniel S. Metlay. "Hazards and Institutional Trustworthiness: Facing a Deficit of Trust." *Public Administration Review* 56.4 (1996): 341–347.

Laurence, D., and B. Wynne. "Transporting Waste in the European Community: A Free Market?" *Environment* 31.6 (1989): 12–17, 34–5.

Lee, Kai N. *Compass and Gyroscope: Integrating Science and Politics for the Environment.* Washington, DC: Island Press, 1993.

Lehman, H. "The Political Economy of Comparative Environmental Policy." *Policy Studies Journal* 20.4 (1992): 719–732.

Lehman, J. P. *Hazardous Waste Disposal.* New York: Plenum Press, 1987.

Leonard, Hugh J. "Confronting Industrial Pollution in Rapidly Industrializing Countries: Myths, Pitfalls, and Opportunities." *Ecology Law Quarterly* 12 (1985): 779–816.

Leonard, Hugh J., and David Morell. "Emergence of Environmental Concern in Developing Countries." *Stanford Journal of International Law* XVII.2 (1981): 281–313.

Leroy, Jean-Bernard. "Hazardous Waste Management in France." *International Perspectives on Hazardous Waste Management.* Eds. William S. Forester and John H. Skinner. London: Academic Press, 1987.

Lester, J. P., J. L. Franke, A. O. Bowman, and K. W. Kramer. "Hazardous Wastes, Politics, and Public Policy: A Comparative State Analysis." *Western Political Quarterly* 36.2 (1983): 257–283.

Leveque, Francois, ed. *Environmental Policy in Europe.* Cheltenham: Edward Elgar, 1996.

Levidow, Les, Susan Carr, David Wield, and René von Schomberg. "European Biotechnology Regulation: Framing the Risk Assessment of a Herbicide-Tolerant Crop." *Science, Technology, and Human Values* 22.4 (1997): 472–406.

Levy, Marc A. "European Acid Rain: The Power of Tote Board Diplomacy." In *Institutions for the Earth: Sources of Effective International Environmental Protection*. Eds. Peter M. Haas, Robert O. Keohane and Marc A. Levy. Cambridge, MA: MIT Press, 1993.

Leyshon, Andrew. "The Transformation of Regulatory Order: Regulating the Global Economy and Environment." *Geoforum* 23.3 (1992): 249–267.

Liefferink, Duncan, and Mikael Skou Andersen. "Strategies of the 'Green' Member States in EU Environmental Policy-Making." *Journal of European Public Policy* 5.2 (1998): 254–270.

Liefferink, J. D., P. D. Lowe, and A. P. J. Mol, eds. *European Integration and Environmental Policy*. London: Belhaven Press, 1993.

Linder, Stephen H. "The Social and Political (Re)Construction of Risk." In *Flashpoints in Environmental Cooperation: Controversies in Achieving Sustainability*. Eds. Sheldon Kamieniecki, George A. Gonzalez and Robert O. Vos. Albany: SUNY Press, 1997.

Linnerooth, Joanne, and Gary Davis. "Government Responsibility for Risk: The Bavarian and Hessian Hazardous Waste Disposal Systems." In *Risk Management and Hazardous Waste: Implementation and the Dialectics of Credibility*. Ed. Brian Wynne. Berlin: Springer-Verlag, 1987.

Linnerooth, Joanne, and Allen V. Kneese. "Hazardous Waste Management: A West German Approach." *Resources* 96 (1989): 7–10.

Lipman, Zada. "The Convention on the Control of Transboundary Movements and Disposal of Hazardous Wastes and Australia's Waste Management Strategy." *Environmental and Planning Law Journal* 7.4 (1990): 283–293.

Lipschutz, Ronnie D., and Ken Conca, eds. *The State and Social Power in Global Environmental Politics*. New York: Columbia University Press, 1993.

Lipschutz, Ronnie D., and Judith Mayer. *Global Civil Society and Global Environmental Governance*. Albany: SUNY Press, 1996.

Litfin, Karen. "Eco-Regimes: Playing Tug of War with the Nation-State." In *The State and Social Power in Global Environmental Politics*. Eds. Ronnie D. Lipschutz and Ken Conca. New York: Columbia University Press, 1993.

Litvan, David. "Politique des déchets: l'approche du Royaume-Uni." *Économie et Statistique* 290.10 (1995): 81–90.

Lotspeich, Richard. "Comparative Environmental Policy: Market-Type Instruments in Industrialized Capitalist Countries." *Policy Studies Journal* 26.1 (1998): 85–104.

Low, Nicholas, and Brendan Gleeson. *Justice, Society, and Nature: An Exploration of Political Ecology*. London: Routledge, 1998.

Low, Patrick, ed. *International Trade and the Environment*. Vol. 159. Washington, DC: World Bank, 1992.

Lowe, Philip, and Andrew Flynn. "Environmental Politics and Policy in the 1980s." In *The Political Geography of Contemporary Britain*. Ed. John Mohan. London: MacMillan, 1989.

Lowe, Philip, and Stephen Ward, eds. *British Environmental Policy and Europe: Politics and Policy in Transition*. London: Routledge, 1998.

Lowi, Theodore. "American Business, Public Policy, Case Studies, and Political Theory." *World Politics* 16: 4 (1964), pp. 677–715.

Lundqvist, Lennart J. "The Comparative Study of Environmental Politics: From Garbage to Gold?" *International Journal of Environmental Studies* 12 (1978): 89–97.

Lundqvist, Lennart J. *The Hare and the Tortoise: Clean Air Policies in the US and Sweden*. Ann Arbor: University of Michigan Press, 1980.

Lynch, Georgina, and Brian Galligan. "Environmental Policymaking in Australia: The Role of the Courts." In *Federalism and the Environment: Environmental Policymaking in Australia, Canada, and the United States*. Eds. Kenneth M. Holland, F. L. Morton, and Brian Galligan. Westport: Greenwood Press, 1996.

Mädel, Eryl, and Brian Wynne. "Decentralized Regulation and Technical Discretion: The U.K." In *Risk Management and Hazardous Waste: Implementation and the Dialectics of Credibility*. Ed. Brian Wynne. Berlin: Springer-Verlag, 1987.

Majone, Giandomenico. *Regulating Europe*. London: Routledge, 1996.

Maltezou, Somà P., Asit K. Biswas, and Hans Sutter, eds. *Hazardous Waste Management: Selected Papers from an International Expert Workshop Convened by UNIDO in Vienna, 22–26 June, 1987*. London: Tycooly, 1989.

Maluwa, Tiyanjana. "Environment and Development in Africa: An Overview of Basic Problems of Environmental Law and Development." *African Journal of International and Comparative Law* 1.4 (1989): 650–671.

Marr, Katharina. *Environmental Impact Assessment in the United Kingdom and Germany: A Comparison of EIA Practice for Wastewater Treatment Plants*. Aldershot: Ashgate Publishing, 1997.

Marshall, Tim. "The Conditions for Environmentally Intelligent Regional Governance: Reflections from Lower Saxony." *Journal of Environmental Planning and Management* 41.4 (1998): 421–443.

McAdam, Doug, John D. McCarthy, and Mayer N. Zald, eds. *Comparative Perspectives on Social Movements: Political Opportunities, Mobilizing Structures, and Cultural Framings*. Cambridge: Cambridge University Press, 1996.

McCormick, John. "Environmental Politics." In *Developments in British Politics 4*. Eds. Patrick Dunleavy, Andrew Gamble, Ian Holliday, and Gillian Peele. London: MacMillan, 1993.

McDonell, Garan. "Toxic Waste Management in Australia: Why Did Policy Reform Fail?" *Environment* 33.6 (1991), pp. 11–13, 33–39.

McKean, Margaret. *Environmental Protest and Citizens Movements in Japan.* Berkeley: University of California Press, 1981.

McKenzie, F. "Policy Formulation for the Management of Hazardous Wastes." *Journal of Environmental Planning and Management* 37.1 (1994): 87–106.

McNeill. "Policy Issues Concerning Transfrontier Movements of Hazardous Wastes." In *Transfrontier Movements of Hazardous Wastes: Legal and Institutional Aspects.* Ed. Organization for Economic Cooperation and Development. Paris: OECD, 1985.

Meereis, Jürgen, and Axel Hartmann. "Sonderabfall: Breiter Konsens über Behandlungstechnologien durch integrativen Bewertungsansatz." *Müll und Abfall* 12 (1994): 808–815.

Mény, Yves. "France: The Institutionalization of Leadership." In *Political Institutions in Europe.* Ed. Josep M. Colomer. London: Routledge, 1996.

Mertig, Angela G., and Riley E. Dunlap. "Public Approval of Environmental Protection and other New Social Movement Goals in Western Europe and the United States." *International Journal of Public Opinion Research* Summer (1995): 145–156.

Mishima, Akio. *Bitter Sea: The Human Cost of Minamata Disease.* Tokyo: Kosei Publishing 1992.

Mitchell, Ronald, and Thomas Bernauer. "Empirical Research on International Environmental Policy: Designing Qualitative Case Studies." *Journal of Environment and Development* 7.1 (1998): 4–31.

Mitchell, Ronald B. *Intentional Oil Pollution at Sea: Environmental Policy and Treaty Compliance.* Cambridge, MA: MIT Press, 1994.

Moe, Terry M., and Michael Caldwell. "The Institutional Foundations of Democratic Government: A Comparison of Presidential and Parliamentary Systems." *Journal of Institutional and Theoretical Economics* 150.1 (1994): 171–195.

Mol, Arthur P. J. "Ecological Modernization and Institutional Reflexivity: Environmental Reform in the Late Modern Age." *Environmental Politics* 5.2 (1996): 302–323.

Montgomery, Mark A. "Politics and Sustainable Development: Guinea-Bissau and Hazardous Waste Imports." Presented at the Annual Meeting of the American Political Science Association, New York, 1994.

Montgomery, Mark A. "Reassessing the Waste Trade Crisis: What Do We Really Know?" *Journal of Environment and Development* 4.1 (1995): 1–28.

Montgomery, Mark A. "Traveling Toxic Trash: An Analysis of the 1989 Basel Convention." *The Fletcher Forum of World Affairs* 14.2 (1990): 313–26.

Montgomery, Mark A. "Want Not, Waste Not: A Realist Theory of the International Trade in Hazardous Waste." Ph.D. Dissertation, Fletcher School of Law and Diplomacy, Tufts University, 1992.

Morrisette, Peter M. "The Montreal Protocol: Lessons for Formulating Policies for Global Warming." *Policy Studies Journal* 19.2 (1991): 152–161.

Moser, Susanne. "Talk Globally, Walk Locally: The Cross-Scale Influence of Global Change Information on Coastal Zone Management in Maine and Hawai'i." ENRP Discussion Paper E-98-16, Kennedy School of Government, Harvard University, 1998.

Moyers, Bill, and Center for Investigative Reporting. *Global Dumping Ground: The International Traffic in Hazardous Wastes*. Washington, DC: Seven Locks Press, 1990.

Munton, Don, ed. *Hazardous Waste Siting and Democratic Choice*. Washington, DC: Georgetown University Press, 1996a.

Munton, Don. "Siting Hazardous Waste Facilities, Japanese Style." In *Hazardous Waste Siting and Democratic Choice*. Ed. Don Munton. Washington, DC: Georgetown University Press, 1996b.

Naito, Sachito. "Hazardous Waste Management in Japan." In *International Perspectives on Hazardous Waste Management*. Eds. William S. Forester and John H. Skinner. London: Academic Press, 1987.

Newman, Penelope. "Killing Legally with Toxic Wastes: Women and the Environment in the United States." In *Close to Home: Women Reconnect Ecology, Health, and Development Worldwide*. Ed. Vandana Shiva. Philadelphia: New Society Publishers, 1994.

Noll, Roger G., ed. *Regulatory Policy and the Social Sciences*. Berkeley: University of California Press, 1985.

Norris, Pippa. "Are We All Green Now? Public Opinion on Environmentalism in Britain." *Government and Opposition* 31.3 (1997): 320–339.

O'Neill, Kate. "Hazardous Waste Disposal." *Foreign Policy in Focus Briefs* January (1999a): 4:1.

O'Neill, Kate. "International Nuclear Waste Transportation: Flashpoints, Lessons, and Controversies." *Environment* 41.4 (1999b): 12–15, 34–39.

O'Neill, Kate. "Out of the Backyard: The Problems of Hazardous Waste Management at a Global Level." *Journal of Environment and Development* 7.2 (1998a): 138–163.

O'Neill, Kate. "(Not) Getting to 'Go': Recent Experience in International Cooperation over the Management of Spent Nuclear Reactor Fuel." BCSIA Discussion Paper 98-22, Kennedy School of Government, Harvard University, 1998b.

O'Neill, Kate. "Regulations as Arbiters of Risk: Great Britain, Germany, and the Hazardous Waste Trade in Western Europe." *International Studies Quarterly* 41.4 (1997): 687–718.

Organization for Economic Cooperation and Development. *Transfrontier Movements of Hazardous Wastes: 1992–93 Statistics.* Paris: OECD, 1997a.

Organization for Economic Cooperation and Development. *France: OECD Environmental Performance Reviews.* Paris: OECD, 1997b.

Organization for Economic Cooperation and Development. *The Environment Industry: The Washington Meeting.* Paris: OECD, 1996.

Organization for Economic Cooperation and Development. *Transfrontier Movements of Hazardous Wastes, 1991 Statistics.* Paris: OECD, 1994a.

Organization for Economic Cooperation and Development. *United Kingdom: OECD Environmental Performance Reviews.* Paris: OECD, 1994b.

Organization for Economic Cooperation and Development. *Regulatory Cooperation for an Interdependent World.* Paris: OECD, 1994c.

Organization for Economic Cooperation and Development. *Transfrontier Movements of Hazardous Wastes, 1989–90 Statistics.* Paris: OECD, 1993a.

Organization for Economic Cooperation and Development. *Monitoring and Control of Transfrontier Movements of Hazardous Wastes.* OECD Environmental Monographs. Vol. 34. Paris: OECD, 1993b.

Organization for Economic Cooperation and Development. *Germany: OECD Environmental Performance Reviews.* Paris: OECD, 1993c.

Organization for Economic Cooperation and Development. *Japan: OECD Environmental Performance Reviews.* Paris: OECD, 1993d.

Organization for Economic Cooperation and Development. *Transfrontier Movements of Hazardous Wastes.* OECD Study 82. Paris: OECD, 1985.

Pakulski, Jan, Bruce Tranter, and Stephen Crook. "The Dynamics of Environmental Concern in Australia: Concerns, Clusters, and Carriers." *Australian Journal of Political Science* 33.2 (1998): 235–252.

Papadakis, Elim, and Anya Moore. "Environment, Economy, and State." In *State, Economy and Public Policy in Australia.* Eds. Stephen Bell and Brian Head. Melbourne: Oxford University Press, 1994.

Park, Rozelia S. "An Examination of International Environmental Racism through the Lens of Transboundary Movement of Hazardous Wastes." *Indiana Journal of Global Legal Studies* 5.2 (1998): 659–709.

Paterson, Matthew. *Global Warming and Global Politics.* London: Routledge, 1996.

Peattie, Ken, and Anja Ringler. "Management and the Environment in the U.K. and Germany: A Comparison." *European Management Journal* 12.2 (1994): 216–225.

Pecqueur, Michel. "Les déchets industriels et leur traitement en France." *Problemes économiques* 2.278 (1992): 24–30.

Peluso, Nancy Lee. "Coercing Conservation: The Politics of State Resource Control." In *The State and Social Power in Global Environmental Politics.*

Eds. Ronnie Lipschutz and Ken Conca. New York: Columbia University Press, 1993.

Pempel, T. J., and Keiichi Tsunekawa. "Corporatism without Labor? The Japanese Anomaly." In *Trends towards Corporatist Intermediation.* Eds. Philippe C. Schmitter and Gerhard Lehmbruch. London: Sage, 1979.

Penna, David R. "Regulation of the Environment in Traditional Society as a Basis for the Right to a Satisfactory Environment." *Africa Today* 40.1 (1993): 82–92.

Peters, A. R. "Germany." In *Local Government in Liberal Democracies.* Ed. J. A. Chandler. London: Routledge, 1992.

Pflügner, Walter, and Martin Götze. "Shipments of Waste within, into and out of the European Community with a View to the Application of the Proximity and the Self-Sufficiency Principles Concerning the Disposal of Waste." Independent report submitted to the Commission of the European Community, DG XI, 1994.

Pharr, Susan J., and Joseph L. Badaracco. "Coping With Crisis: Environmental Regulation." In *America Versus Japan.* Ed. Thomas K. McCraw. Boston: Harvard Business School Press, 1988.

Piasecki, Bruce, and Janet Brooks. "Government's Aid: The Role of Citizen and Environmental Groups in Europe." In *America's Future in Toxic Waste Management: Lessons from Europe.* Eds. Bruce Piasecki and Gary Davis. New York: Quorum Books, 1987.

Piasecki, Bruce, and Gary Davis, eds. *America's Future in Toxic Waste Management: Lessons from Europe.* New York: Quorum Books, 1987.

Piasecki, Bruce, and Gary Davis. "A Grand Tour of Europe's Hazardous Waste Facilities." *Technology Review* 87.5 (1984): 20–29.

Porter, Gareth, and Janet Welsh Brown. *Global Environmental Politics.* 2nd ed. Boulder, CO: Westview, 1995.

Porter, Martin. "Waste Management." In *British Environmental Policy and Europe: Politics and Policy in Transition.* Eds. Philip Lowe and Stephen Ward. London: Routledge, 1998.

Portney, Kent. "The Role of Economic Factors in Lay Perceptions of Risk." In *Dimensions of Hazardous Waste Politics and Policy.* Eds. Charles E. Davis and James P. Lester. New York: Greenwood Press, 1988.

Postel, Sandra. "Defusing the Toxics Threat: Controlling Pesticides and Industrial Waste." *World Watch Papers* Number 79 (1987).

Potter, David. "Assessing Japan's Environmental Aid Policy." *Pacific Affairs* 67.2 (1994): 200–215.

Powell, Douglas, and William Leiss. *Mad Cows and Mother's Milk: The Perils of Poor Risk Communication.* Montreal: McGill-Queen's University Press, 1997.

Prendiville, Brendan. *Environmental Politics in France*. Boulder, CO: Westview, 1994.

Pridham, Geoffrey. "National Environmental Policy in the European Framework: Spain, Greece, and Italy in Comparison." In *Protecting the Periphery: Environmental Policy in the Peripheral Regions of the European Union*. Eds. Susan Baker, Kay Milton, and Steven Yearley. Ilford: Frank Cass, 1994.

Princen, Thomas, and Matthias Finger. *Environmental NGOs in World Politics: Linking the Local and the Global*. London: Routledge, 1994.

Puckett, Jim. "Disposing of the Waste Trade: Closing the Recycling Loophole." *The Ecologist* 24.2 (1994): 53–8.

Rabe, Barry G., William C. Gunderson, and Peter T. Harbage. "Alternatives to NIMBY Gridlock: Voluntary Approaches to Radioactive Waste Facility Siting in Canada and the United States." In *Hazardous Waste Siting and Democratic Choice*. Ed. Don Munton. Washington, DC: Georgetown University Press, 1996.

Rabe, Barry. "Power to the States: The Promise and Pitfalls of Decentralization." In *Environmental Policy in the 1990s*. 3rd ed. Eds. Norman J. Vig and Michael E. Kraft. Washington, DC: Congressional Quarterly Press, 1997.

Rahm, Dianne. "Superfund and the Politics of U.S. Hazardous Waste Policy." *Environmental Politics* 7.4 (1998): 75–91.

Randlesome, Collin. *Business Culture in Germany: Portrait of a Powerhouse*. Oxford: Butterworth-Heinemann, 1994.

Raustiala, Kal. "Domestic Institutions and International Regulatory Cooperation: Comparative Responses to the Convention on Biological Diversity." *World Politics* 49.4 (1997): 482–509.

Rawcliffe, Peter. *Environmental Pressure Groups in Transition*. Manchester: Manchester University Press, 1998.

Redclift, Michael, and Ted Benton, eds. *Social Theory and the Global Environment*. London: Routledge, 1994.

Reed, Steven R. *Japanese Prefectures and Policymaking*. Pittsburgh: Pittsburgh University Press, 1986.

Rehbinder, E. "Vorsorgeprinzip, Umweltrecht und Präventive Umweltpolitik." *Präventive Umweltpolitik*. Ed. U. Simonis. Frankfurt: Campus Verlag, 1988.

Reich, M. R. "Mobilizing for Environmental Policy in Italy and Japan." *Comparative Politics* 16 (1984): 379–402.

Renn, Ortwin, William J. Burns, Jeanne X. Kasperson, Roger E. Kasperson, and Paul Slovic. "The Social Amplification of Risk: Theoretical Foundations and Empirical Applications." *Journal of Social Issues* 4 (1992): 137–160.

Rethmann, Norbert. "Konsequenzen der TA Sonderabfall für das Innovationsverhalten der Entsorgungsunternehmen." In *Unternehmung und ökologische Umwelt*. Ed. Gerd Rainer Wagner. Munich: Verlag Franz Vahlen, 1990.

Rhodes, R. A. W. *Beyond Westminster and Whitehall: The Sub-Central Governments of Britain.* London: Unwin Hyman, 1988.

Rhodes, R. A. W. "The Hollowing Out of the State: the Changing Nature of the Public Service in Britain." *Political Quarterly* 65.2 (1994): 138–151.

Rhodes, R. A. W. *Reinventing Whitehall, 1979–95?* Paper presented at the American Political Science Association Annual Meeting, San Francisco, 1996.

Ribot, Jesse C. "Market-State Relations and Environmental Policy: Limits of State Capacity in Senegal." In *The State and Social Power in Global Environmental Politics.* Eds. Ronnie Lipschutz and Ken Conca. New York: Columbia University Press, 1993.

Richardson, Derek, and Chris Rootes, eds. *The Green Challenge: The Development of Green Parties in Europe.* London: Routledge, 1995.

Richardson, Jeremy, ed. *Policy Styles in Western Europe.* London: George Allen and Unwin, 1982.

Richardson, Jeremy, and Andrew Jordan. *Governing Under Pressure: The Policy Process in a Post-Parliamentary Democracy.* Oxford: Martin Robertson, 1979.

Robinson, Brian. "Hazardous Waste Management in Victoria, Australia." *Waste Management and Research* 8 (1990): 99–103.

Robinson, Tim. "The Environment and Business Regulation." *Government and Business Relations in Australia.* Ed. Randal G. Stewart. Sydney: Allen & Unwin, 1994.

Rootes, Chris. "Britain: Greens in a Cold Climate." In *The Green Challenge: The Development of Green Parties in Europe.* Eds. Dick Richardson and Chris Rootes. London: Routledge, 1995.

Rose, Chris. *The Dirty Man of Europe: The Great British Pollution Scandal.* London: Simon & Schuster, 1990.

Rose-Ackerman, Susan. *Controlling Environmental Policy: The Limits of Public Law in Germany and the United States.* New Haven: Yale University Press, 1995.

Rosenbaum, Walter A. "Regulation at Risk: The Controversial Politics and Science of Risk Assessment." In *Flashpoints in Environmental Cooperation: Controversies in Achieving Sustainability.* Eds. Sheldon Kamieniecki, George A. Gonzalez and Robert O. Vos. Albany: SUNY Press, 1997.

Rosencranz, Armin, and Christopher L. Eldridge. "Hazardous Wastes: Basel after Rio." *Environmental Policy and Law* 22.5/6 (1992): 318–322.

Rüdig, Wolfgang. "Comparing Environmental Policies: The Anglo-German Conundrum." *Environmental Politics* 2.3 (1993): 504–508.

Sabatier, Paul A., and Hank C. Jenkins-Smith, eds. *Policy Change and Learning: An Advocacy Coalitions Approach.* Boulder, CO: Westview Press, 1993.

Sand, Peter H. *Lessons Learned in Global Environmental Governance.* Washington, DC: World Resources Institute, 1990.

Saunders, Cheryl. "The Constitutional Division of Powers with Respect to the Environment in Australia." In *Federalism and the Environment: Environmental Policymaking in Australia, Canada, and the United States*. Eds. Kenneth M. Holland, F. L. Morton, and Brian Galigan. Westport: Greenwood Press, 1996.

Sbragia, Alberta. *The European Community and Implementation: Environmental Policy in Comparative Perspective*, 1991. Presented at the Annual Meeting of the American Political Science Association, Washington, D.C., 1991.

Scheberle, Denise. *Federalism and Environmental Policy: Trust and the Politics of Implementation*. Washington, DC: Georgetown University Press, 1997.

Schmidt, Manfred G. "Germany: The Grand Coalition State." In *Political Institutions in Europe*. Ed. Josep M. Colomer. London: Routledge, 1996.

Schmitt-Tegge, Jan. "Incineration of Hazardous and Municipal Waste: The Situation in Germany in 1993." *Waste Management and Research* 12 (1994): 441–453.

Schreurs, Miranda A. "Domestic Institutions and International Environmental Agendas in Japan and Germany." In *The Internationalization of Environmental Protection*. Eds. Miranda A. Schreurs and Elizabeth C. Economy. Cambridge: Cambridge University Press, 1997.

Schreurs, Miranda A., and Elizabeth C. Economy, eds. *The Internationalization of Environmental Protection*. Cambridge: Cambridge University Press, 1997.

Scruggs, Lyle A. "Institutions and Environmental Performance in Seventeen Western Democracies." *British Journal of Political Science* 29.1 (1999): 1–31.

Seager, Joni. *Earth Follies: Feminism, Politics and the Environment*. London: Earthscan, 1993.

Shull, Tad. "Green Politics and Political Mobilization: Contradictions of Direct Democracy." In *The Mitterrand Era: Political Mobilization and Policy Alternatives in France*. Ed. Anthony Daley. New York: MacMillan, 1996.

Siegmann, Heinrich. *The Conflicts between Labor and Environmentalism in the Federal Republic of Germany and the United States*. New York: St. Martin's Press, 1985.

Sierig, Gerhard. "Hazardous Waste Management in the Federal Republic of Germany." In *International Perspectives on Hazardous Waste Management*. Eds. William S. Forester and John H. Skinner. London: Academic Press, 1987.

Sikkink, Kathryn. *Ideas and Institutions: Developmentalism in Argentina and Brazil*. Ithaca: Cornell University Press, 1991.

Simmons, Beth A. "Compliance with International Agreements." *Annual Review of Political Science* 1 (1998): 75–93.

Simon, Joel. *Endangered Mexico: An Environment on the Edge*. San Francisco: Sierra Club Books, 1997.

Singh, Jang B., and V. C. Lakahan. "Business Ethics and the International Trade in Hazardous Wastes." *Journal of Business Ethics* 8 (1989): 889–899.

Skea, Jim, and Adrian Smith. "Integrating Pollution Control." In *British Environmental Policy and Europe: Politics and Policy in Transition*. Eds. Philip Lowe and Stephen Ward. London: Routledge, 1998.

Souet, Patrick. "The French National Plan for Environment: A New Impulse in French Waste Management Policy." *Waste Management and Research* 11.2 (1993): 171–176.

Sprinz, Detlef F. "Measuring the Effectiveness of International Environmental Regimes." Presented at the Annual Meeting of the International Studies Association, San Diego, CA., 1996.

Sprinz, Detlev, and Tapani Vaahtoranta. "The Interest-Based Explanation of International Environmental Policy." *International Organization* 48.1 (1994): 77–105.

Stanners, David, and Philippe Bourdeau, eds. *Europe's Environment: The Dobris Assessment*. Copenhagen: European Environment Agency, 1995.

Stark, David. "Path Dependency and Privatization Strategies in Eastern Europe." *East European Politics and Society* 6.1 (1992): 17–54.

Stede, Birgit. "Abfall in der Kreislaufwirtschaft." *Müll und Abfall* 3 (1996): 141–152.

Steinmo, Sven. "Political Institutions and Tax Policy in the United States, Sweden, and Great Britain." *World Politics* 41.4 (1989): 500–35.

Steinmo, Sven, Kathleen Thelen, and F. Longstreth, eds. *Structuring Politics: Historical Institutionalism in Comparative Analysis*. Cambridge: Cambridge University Press, 1992.

Stevens, Anne. *The Government and Politics of France*. 2nd ed. New York: St. Martin's Press, 1996.

Stevis, Dimitris, Valerie J. Assetto, and Stephen P. Mumme. "International Environmental Politics: A Review of the Literature." In *Environmental Politics and Policy: Theories and Evidence*. Ed. James P. Lester. Durham: Duke University Press, 1989.

Stiglitz, Joseph. "Principal and Agent." In *The New Palgrave: A Dictionary of Economics*. Vol. 3. Eds. John Eatwell, Murray Milgate, and Peter Newman. London: Macmillan, 1987.

Stockwin, J. A. A. *Japan: Divided Politics in a Growth Economy*. New York: Weidenfeld and Nicolson, 1982.

Strohm, Laura. "The Environmental Politics of the International Waste Trade." *Journal of Environment and Development* 2.2 (1993): 129–53.

Susskind, Lawrence E. *Environmental Diplomacy: Negotiating More Effective Global Environmental Agreements*. New York: Oxford University Press, 1994.

Susskind, Lawrence E., E. Siskind, and J. W. Breslin, eds. *Nine Case Studies in International Environmental Negotiation*. Cambridge, MA: MIT-Harvard Public Disputes Program, 1990.

Swift, Jonathan, "A Digression Concerning Criticks." In *Tale of a Tub*, London: Nonesuch Press (1696, reprinted 1949).

Tarrow, Sidney. *Power in Movement: Social Movements and Contentious Politics*. 2nd ed. Cambridge: Cambridge University Press, 1998.

Taylor, Bron, ed. *Ecological Resistance Movements: The Global Emergence of Radical and Popular Environmentalism*. Albany: SUNY Press, 1995.

Thelen, Kathleen, and Sven Steinmo. "Historical Institutionalism in Comparative Perspective." In *Structuring Politics: Historical Institutionalism in Comparative Perspective*. Eds. Sven Steinmo, Kathleen Thelen, and Frank Longstreth. Cambridge: Cambridge University Press, 1992.

Third World Network. *Toxic Terror: The Dumping of Hazardous Wastes in the Third World*. Penang: Third World Network, 1989.

Thompson, Michael. "Blood, Sweat and Tears." *Waste Management and Research* 12.3 (1994): 199–206.

Thompson, Peter, and Laura A. Strohm. "Trade and Environmental Quality: A Review of the Evidence." *Journal of Environment and Development* 5.4 (1996): 363–388.

Tolba, Mostafa K., and Osama A. El-Kholy, eds. *The World Environment 1972–1992: Two Decades of Challenge*. London: Chapman and Hall, 1992.

Tsuru, Shigeto. "History of Pollution Control Policy." In *Environmental Policy in Japan*. Eds. Shigeto Tsuru and Helmut Weidner. Berlin: Edition Sigma Bohn Verlag, 1989.

Ui, Jun. "Anti-Pollution Movements and Other Grass-Roots Organizations." In *Environmental Policy in Japan*. Eds. Shigeto Tsuru and Helmut Weidner. Berlin: Edition Sigma Bohn Verlag, 1989.

Ui, Jun, ed. *Industrial Pollution in Japan*. Tokyo: United Nations University Press, 1992.

Umwelt, various issues.

Unger, Brigitte, and Frans van Warden, eds. *Convergence or Diversity? Internationalization and Economic Policy Response*. Aldershot: Avebury, 1995.

United Church of Christ Commission for Racial Justice. "Toxic Wastes and Race in the United States: A National Report on the Racial and Socioeconomic Characteristics of Communities with Hazardous Waste Sites." New York: Authur, 1987.

Vallette, Jim, and Heather Spalding. *The International Trade in Hazardous Wastes: A Greenpeace Inventory*. 5th ed. Washington, DC: Greenpeace International Waste Trade Project, 1990.

van Waarden, Frans. "Persistence of National Policy Styles: A Study of their Institutional Foundations." In *Convergence or Diversity? Internationalization and Economic Policy Response*. Eds. Brigitte Unger and Frans van Waarden. Aldershot: Avebury, 1995.

VanDeveer, Stacy D. "Normative Force: The State, Transnational Norms, and International Environmental Regimes." Ph.D. dissertation. University of Maryland, 1997.

Victor, David G., Kal Raustiala, and Eugene B. Skolnikoff, eds. *The Implementation and Effectiveness of International Environmental Commitments: Theory and Practice.* Cambridge, MA: MIT Press, 1998.

Vogel, David. "Consumer Protection and Protectionism in Japan." *Journal of Japanese Studies* 18.1 (1992): 119–154.

Vogel, David. "International Trade and Environmental Regulation." In *Environmental Policy in the 1990s: Reform or Reaction.* 3rd ed. Eds. Norman J. Vig and Michael E. Kraft. Washington, DC: Congressional Quarterly Press, 1997.

Vogel, David. *National Styles of Regulation: Environmental Policy in Great Britain and the United States.* Ithaca: Cornell University Press, 1986.

Vogel, David. *Trading Up: Consumer and Environmental Regulation in a Global Economy.* Cambridge, MA: Harvard University Press, 1995.

Vogel, David, and Veronica Kun. "The Comparative Study of Environmental Policy: A Review of the Literature." In *Comparative Policy Research: Learning from Experience.* Eds. Meinoff Dierkes, Hans N. Weiler, and Ariane Berthoin Antal. Berlin: WZB-Publications, 1987.

Vogler, John, and Mark F. Imber, eds. *The Environment and International Relations.* London: Routledge, 1996.

Voisey, Heather, and Tim O'Riordan. "Governing Institutions for Sustainable Development: The United Kingdom's National Level Approach." *Environmental Politics* 6.1 (1997): 24–53.

Walker, K. J. *The Political Economy of Environmental Policy: An Australian Introduction.* Sydney: University of New South Wales Press, 1994.

Wallace, David. *Environmental Policy and Industrial Innovation: Strategies in Europe, the U.S.A., and Japan.* London: Earthscan, 1995.

Wapner, Paul. *Environmental Activism and World Civic Politics.* Albany: SUNY Press, 1996.

Wapner, Paul. "Politics Beyond the State: Environmental Activism and World Civic Politics." *World Politics* 47.3 (1995): 311–340.

Ward, Stephen. "Thinking Global, Acting Local? British Local Authorities and Their Environmental Plans." *Environmental Politics* 2.3 (1993): 453–78.

Weale, Albert. "Ecological Modernization and the Integration of European Environmental Policy." In *European Integration and Environmental Policy.* Eds. J. D. Liefferink, P. D. Lowe, and A. P. J. Mol. London: Belhaven Press, 1993.

Weale, Albert. "Environmental Regulation and Administrative Reform in Britain." In *Regulating Europe.* Ed. Giandomenico Majone. London: Routledge, 1996.

Weale, Albert. "The Kaleidoscopic Competition of European Environmental Regulation." *European Business Journal* 7.4 (1995): 19–25.

Weale, Albert. *The New Politics of Pollution*. Manchester: Manchester University Press, 1992a.

Weale, Albert. "Vorsprung durch Technik? The Politics of German Environmental Regulation." In *The Politics of German Regulation*. Ed. K. Dyson. Aldershot: Dartmouth Publishing Company, 1992b.

Weale, Albert, Timothy O'Riordan, and L. Kramme. *Pollution in the Round: Change and Choice in Environmental Regulation in Britain and West Germany*. London: Anglo-German Foundation, 1991.

Weale, Albert, Geoffrey Pridham, Anthea Williams, and Martin Porter. "Environmental Administration in Six European Countries: Secular Convergence or National Distinctiveness?" *Public Administration* 74.2 (1996): 255–274.

Weidner, Helmut. "An Administrative Compensation System for Pollution-Related Health Damages." In *Environmental Policy in Japan*. Eds. Shigeto Tsuru and Helmut Weidner. Berlin: Edition Sigma Bohn Verlag, 1989a, pp. 139–165.

Weidner, Helmut. "Japanese Environmental Policy in an International Perspective: Lessons for a Preventive Approach." In *Environmental Policy in Japan*. Eds. Shigeto Tsuru and Helmut Weidner. Berlin: Edition Sigma Bohn Verlag, 1989b, pp. 479–552.

Weinthal, Erika. "Making or Breaking the State? Building Institutions for Regional Cooperation in the Aral Sea Basin." Ph.D. dissertation. Columbia University, 1998.

Weiss, Edith Brown, and Harold K. Jacobson, eds. *Engaging Countries: Strengthening Compliance with International Environmental Accords*. Cambridge, MA: MIT Press, 1998.

Wildavsky, Aaron, and Karl Dake. "Theories of Risk Perception: Who Fears What and Why?" *Daedelus* 119.4 (1990): 41–60.

Williamson, John. "Local Government as a Context for Policy." In *Policy-Making in Britain: An Introduction*. Ed. Maurice Mullard. London: Routledge, 1995.

Wilson, David C. "Hazardous Waste Management in the United Kingdom." In *International Perspectives on Hazardous Waste Management*. Eds. William S. Forester and John H. Skinner. London: Academic Press, 1987.

Wilson, David C., and Fritz Balkau. "Adapting Hazardous Waste Management to the Needs of Developing Countries: An Overview and Guide to Action." *Waste Management and Research* 8 (1990): 87–97.

Wimmer, Frank, and Heiko Wahl. "Value Change and Environmental Awareness in Germany." In *The Green Agenda: Environmental Politics and Policy in Germany*. Eds. Ingolfur Blühdorn, Frank Krause, and Thomas Scharf. Keele: Keele University Press, 1995.

Winter, Michael. *Rural Politics: Policies for Agriculture, Forestry, and the Environment*. London: Routledge, 1996.

Wintle, Michael, and Rachel Reeve, eds. *Rhetoric and Reality in Environmental Policy: The Case of the Netherlands in Comparison with Britain*. Aldershot: Avebury Studies in Green Research, 1994.

Worcester, Robert. "Public Opinion and the Environment." In *Greening the Millennium? The New Politics of the Environment*. Ed. Michael Jacobs. Oxford: Blackwell Publishers, 1997.

World Resources Institute. "Managing Hazardous Wastes: The Unmet Challenge." In *World Resources, 1987*. New York: Basic Books, 1987.

World Resources Institute. *World Resources 1994–5: A Guide to the Global Environment*. New York: Oxford University Press, 1994.

Wright, Vincent. "The Administrative Machine: Old Problems and New Dilemmas." In *Developments in French Politics*. Rev. ed. Eds. Peter A. Hall, Jack Hayward, and Howard Machin. London: MacMillan, 1994.

Wynne, Brian, ed. *Risk Management and Hazardous Waste: Implementation and the Dialectics of Credibility*. Berlin: Springer-Verlag, 1987.

Wynne, Brian. "The Toxic Waste Trade: International Regulatory Issues and Options." *Third World Quarterly* 11.3 (1989): 120–46.

Yakowitz, H. "Waste Management: What Now? What Next? An Overview of Policies and Practices in the OECD Area." *Resources, Conservation, and Recycling* 8 (1993): 131–178.

Yarrow, George, and John Vickers. *Privatization: An Economic Analysis*. Cambridge, MA: MIT Press, 1988.

Young, Oran R. "Global Environmental Change and International Governance." *Millennium* 19.3 (1990): 337–346.

Young, Oran R. *International Cooperation: International Regimes for Natural Resources and the Environment*. Ithaca: Cornell University Press, 1989.

Young, Oran R. *International Governance: Protecting the Environment in a Stateless Society*. Ithaca: Cornell University Press, 1994.

Zaelke, Durwood, Paul Orbuch, and Robert F. Housman, eds. *Trade and the Environment: Law, Economics, and Policy*. Washington, DC: Island Press, 1993.

Zito, Anthony R. "The Role of Technical Expertise and Institutional Innovation in European Hazardous Waste Policy." Presented at the Annual Meeting of the American Political Science Association, New York, 1994.

Zürn, Michael. "The Rise of International Environmental Politics: A Review of the Current Research." *World Politics* 50.4 (1998): 617–649.

Zysman, John. *Governments, Markets, and Growth: Financial Systems and the Politics of Industrial Change*. Ithaca: Cornell University Press, 1983.

Index

Abfall. See Germany
Abromeit, Heidrun, 122, 123
access, policy process, 61. *See also* individual countries
　policy access, 61–62
　open/inclusive styles, 62
　closed/exclusive styles, 62
Adams, Carolyn Teich, 156
Africa, 1, 37, 40, 41, 209, 210
Agenda, 21, 260
Aguilar, Susan, 17
Alkali Inspectorate, 78
Australia, 2, 5, 7, 8, 21, 22, 23, 34, 35, 53, 69, 70, 147, 160–169, 182, 183, 185, 188, 189, 192, 194, 210, 214
　Aboriginal and Torres Strait Islander Commission, 163
　alternative waste disposal, 167
　Australian Conservation Foundation, 167, 255
　Base Catalyzed Dechlorination, 255
　Biological Control Act—1984, 163
　bioremediation, 255
　Coode Island, 167
　Corowa, 166–168
　Democrats, 165
　Department of Environment, Sports, and Territories, 163
　Department of Primary Industries and Energy, 163
　Ecologic, 255
　Environmental Protection (Impact of Proposals) Act, 253
　Environmental Protection Agency, 163
　exports, waste, 148, 222–226
　France, 160
　less developed countries, ban, 168
　United Kingdom, 160
　Franklin River Dam, 162, 163
　Green Party, 165, 255
　Hawke, Bob, 162, 163
　Hazardous Waste (Regulation of Exports and Imports) Act—1989, 164
　High Court, 162
　House of Representatives, 165
　incineration, 166, 167
　Intergovernmental Agreement on the Environment, 163
　Labor Party, 162, 165
　landfills, 160, 167, 210 (*see also* individual countries)
　Liberal Party, 169, 254
　Melbourne, 163, 167
　New South Wales, 160, 161, 163, 166, 192, 253
　ocean dumping, 160
　Ozone Protection Act—1989, 164
　recycling, 168
　regulation
　　structure of, 161–164
　　style, 164–166

Australia (cont.)
  Senate, 165
  Sydney, 160, 163, 167, 192
  Tasmania, 162, 165, 255
  Victoria, 161, 163, 166, 192, 253
  waste, intractable, 253
  Whitlam, Gough, 162
  Wildlife Protection (Regulation of Exports and Imports) Act—1982, 164
Austria, 115

Badaracco, Joseph L., 170, 177
Baldwin, Paul, 14
Bamako Convention, 41
Barcelona Convention. *See* Izmir Protocol
Basel Convention on the Transfrontier Movement of Hazardous Wastes and Their Disposal, 3, 4, 9, 23, 27, 40, 41, 42–43, 46, 65, 68, 77, 111, 115, 168, 169, 194, 211, 214
  Annex VII, 43, 168
  environmentally sound manner, 42
  prior informed consent, 42
  Technical Working Groups, 212, 213
BASF, 244
Beck, Ulrich, 32
Beecher, Norman, 172
Belgium, 31, 143
  waste exports to the United Kingdom, 78
Bertossi, Paolo, 212
best practicable environmental option, 90, 239
Best Practical Means criterion, 90, 239
Bhopal, 225
Blair, Tony, 77, 239
Blühdorn, Ingolfur, 129
Bodiguel, Maryvonne, 153
Bottomley, Virginia, 98
Boyd, Richard, 176
Brent Spar Oil Rig, 45, 202
Brickman, Ronald, 17, 134, 155

Britain. *See* Great Britain
Browning Ferris Industries, 92
Brundtland Commission Report on Sustainable Development, 204
Brusco, Sebastiano, 212
BSE. *See* Great Britain
Buller, Henry, 153, 160
Bureau of International Recycling, 227

cadmium, 27
Caldwell, Michael, 17, 18, 107
California effect. *See* regulatory convergence
Cambodia, 209
Canada, 36
Caribbean, 1, 40
Carson, Rachel, 32
Chernobyl, 123
Chile, 35
China, 210, 213
chlorofluorocarbons, 128
Cleanaway, 92, 97, 99
codisposal, 210. *See also* Great Britain
Compagnie Générale des Eaux, 92, 251
comparative advantage in waste disposal
  as explanation, 67, 68–69, 186–188
  capacity, 67
  Germany, 117–120
  role of costs, 119–120
  Great Britain, 101–108
  technology, 67–69
comparative environmental policy, 15
Conca, Ken, 14
Copeland, Brian, 19
Cottica, Alberto, 212
Crook, Stephen, 166

Davis, Gary, 125
Defregger, Franz, 119
Delaware effect. *See* regulatory convergence

Denmark, 31
DG XI. *See* European Community
dioxins, 3, 44
Dunlap, Riley, 16

Eastern Europe, 37, 203
East Germany. *See* German Democratic Republic
ecological modernization, 204–207
Economic Nationalist Explanation, 66, 186–187
  Great Britain, 100, 108
Economy, Elizabeth C., 14
ECOTEC, 242
EMAS, 211
ENDS Report, 72, 78, 99
environmental justice, 37
environmental policy, comparative, 195–197
environmental politics, international, 193–195
Environmental Services Association, 92, 93, 260
Ethiopia, 209
European Commission, 45
European Community, 35, 36
  DG XI, 44, 201
  Directive on the Supervision and Control within the European Community of the Transfrontier Shipments of Hazardous Waste, 44–45
  Packaging Waste Directive, 45
  Waste Directive, 44
European Environment Agency, 201, 259
European Union, 4, 18, 20, 23, 41, 43–46, 65, 69, 71, 75, 110, 144, 146, 197, 203, 210, 212, 213, 215
  Integrated Pollution Prevention and Control Directive—1996, 200, 236
  Single European Act, 43
  Treaty on European Union, 44
externalities, 33

Forester, William S., 17
France, 2, 5, 7, 8, 21, 22, 23, 31, 41, 44, 53, 69, 70, 143, 147, 182, 185, 188, 194, 197, 203, 205, 214
  Compagnie Générale des Eaux, 155
  Déchets spéciaux, 154
  Decree, 92–98, 158
  directorates of industry, research, and the environment, 251
  Environment Ministry, 153, 159
  Fédération Nationale des Sociétés pour la Protection de la Nature, 157
  imports, hazardous waste, 148–150, 251
  hospital wastes, from Germany, 158
  incineration, 154, 159, 252
  landfills, 154, 159, 252
  Les Verts, 157, 158, 252
  Lyonnaise des Eaux, 155
  Marseilles, 156
  National Agency for the Recovery and Elimination of Waste (ANRED), 153
  National Environmental Policy Plan, 149
  National Front, 157
  regulation
    structure of environmental, 150–154
    style of environmental, 155–157
  Rhone-Alpes, 158
  stakeholder relations
    comparison with Britain and Germany, 158
    center-local government relations, 151–152
    institutional constraints and risk acceptance, 157–159, 197
    waste management industry, 154–155
  Waste Management Plan—1992, 154
Friends of the Earth, 6, 95, 96, 241

Gallup Health of the Planet Survey, 16
General Agreement on Tariffs and Trade, 36
Générale des Eaux, 35, 212
genetically modified organisms, 33
German Democratic Republic, 113, 115, 124, 243
  Schönberg waste dump, 243
Germany, 1, 2, 5, 6, 7, 21, 22, 23, 41, 45, 53, 60, 65, 66, 69, 70, 95, 96, 182, 185, 188, 190, 191,193, 197, 200, 204, 214, 226
  Abfall, 72
  Abfallbeseitigungsgesetz (Waste Disposal Law), 125
  access, policy process, 135–141
  BASF, 118
  Bayer, 118
  Bundesministerium für Umwelt, Naturschutz und Reaktorsicherheit (BMU), 123, 128
  Bundesrat, 121, 130
  Bundestag, 121, 130, 137
  Bundesverfassungsgericht (*see* Germany, Federal Constitutional Court)
  Christian Democratic Union, 130, 247
  Clean Air Maintenance Law—1959, 123
  closed circle economy (*see* Germany, Kreislaufwirtschaft)
  comparative advantage (*see* comparative advantage in waste disposal, Germany)
  Conference of Environment Ministers, 124
  Council of Economic Experts, 128–129
  DGB, 249
  disposal capacity, 118–119
  disposal technology, 117–118
  Duales System Deutschland, 250
  East Germany, former, 144, 145
  Environmental Cabinet, 124
  Federal Constitutional Court, 131
  Federal Democratic Party, 130, 137, 247
  Federal Environment Agency, 118, 119, 123
  Federal Water Resources Act—1957, 123
  Genossenschaften, 122
  Green Party, 7, 130, 136–138, 132, 140, 206, 248
  Grundgesetz, 123
  Hamburg, 117
  Hoechst, 118
  incineration, 118, 119
  institutional perspective, 134–135
    comparison with Great Britain, 135
  Kooperationsprinzip, 128
  Kreislaufwirtschaft (closed circle economy), 125, 133, 144, 146, 192
  Land Environmental Ministers, 124
  Länder, 84, 121, 122, 124, 125, 130, 131, 139, 143
  Baden-Württemburg, 116–117, 136, 243
  Bavaria, 125, 244, 245
  Bremen, 139
  division of powers with federal government, 123–124
  Hessen, 119, 124, 139, 244, 245
  Lower Saxony, 126, 139
  Neues Bundesländer, imports from other Länder, 139
  North Rhine-Westfalia, 125, 245
  Schleswig-Holstein, 139
  landfills, 119
  Management of Special Wastes in Bavaria (GSB), 125
  Mittelfranken Cooperative for Special Waste Management, 125
  Müll, 72
  nongovernmental organizations, 132, 248
  German Alliance for Environment and Nature Conservation, 248

German League for the Protection of Nature, 248
Greenpeace Germany, 248
Robin Wood, 248
World Wide Fund for Nature Germany, 248
Packaging Ordinance—1990, 133, 145, 250
policy formation, process of, 127–129, 138–141
  comparison with Great Britain, 138–139
policy implementation, environmental, mode of, Germany, 127, 132–134
  comparison with Great Britain, 134
polluter pays principle, 132, 133, 143
precautionary principle, 131, 132, 133, 143, 191, 247
principal-agent dilemmas, 141–142
regulation
  structure of environmental, 121–127, 141–142
  style of environmental, 127–132
  system of environmental, 120–121
Social Democratic Party, 130, 137, 138, 247
Staatlichkeit, 140
Stand der Technik (*see* Germany, very best technology)
Umweltbundesamt (UBA), 123, 129
Verursacherprinzip (*see* Germany, polluter pays principle)
very best technology, 133, 143
Vorsorgeprinzip (*see* Germany, precautionary principle)
Waldsterben, 120
waste exports
  to France, 117, 136, 144
  illegal, 117, 225
  to United Kingdom, 78
waste imports, 115
waste minimization, 214
waste, special, 117, 245
Wirtschaftswunder, 120, 122

Ghana, 230
Götze, Martin, 139
Great Britain, 1, 2, 5, 6, 7, 8, 10, 21, 22, 23, 37, 40, 53, 60, 65, 69, 70, 75, 182, 188, 190, 191, 193, 194, 197, 202, 203, 211, 214
access, policy process, 88–90
Bedfordshire, 87
BSE, 33, 202
Cheshire, 87
Clean Air Act of 1956, 81, 96
codisposal, 103
comparative advantage (*see* comparative advantage in waste disposal, Great Britain)
Conservative Party, 87, 89
Control of Pollution Act of 1974, 81, 82, 86
Control of Pollution (Special Waste) Regulations of 1980, 82–83, 86
Daily Mirror, 94
decentralized structure, 75, 85–88
Department of the Environment, Transport, and the Regions, 77, 83
Department of Industry, 83, 96
Deposit of Poisonous Wastes Act of 1972, 86
Deposit of Toxic Wastes Act of 1972, 82
disposal capacity, 103–106
Environment Agency, 72, 109
Environmental Protection Act of 1990, 77, 83
Environmental Services Association, 92, 93, 260
Environmental White Paper of 1980, 99
Essex, 87
Green Party, 89, 95, 97
Hazardous Waste Inspectorate, 86, 89, 101, 103, 106, 108
Herald of Free Enterprise, 98
Her Majesty's Inspectorate of Pollution, 84
incineration capacity, 105

Great Britain (cont.)
Industrial Air Pollution Inspectorate, 83
Integrated Pollution Control, System of, 82, 198, 200, 231, 236
Karin B., 97, 98
Labour Party, 89, 110
landfills, 103, 105, 242
landfill tax, 109
Local Government Planning and Land Act of 1980, 85
mad cow disease, 96, 242
Manchester, 87
Ministry for Agriculture, Fisheries, and Food, 83, 96
National Association of Waste Disposal Contractors (*see* Environmental Services Association)
National Rivers Authority, 84
National Transfrontier Shipment Service, 109
National Waste Management Planning Guide and the Management Plan for Exports and Imports of Waste, 109
Policy implementation, environmental, mode of, 75–76, 90–91
Radioactive Substances Inspectorate, 83
regulation
  structure of environmental, 83–88
  style of environmental, 88–91
  system of environmental, 78–83
Report of the House of Lords Select Commission on Science and Technology, on Hazardous Waste Disposal, 85
Report of the Royal Commission on Environmental Pollution, 82
special wastes
  definition, 109–110, 236
  generation, 105
  imports, 77

Town and Country Planning Act—1947, 237
Waste Collection Authorities, 86
Waste Disposal Authorities, 86, 87
waste disposal methods, 102–103
waste imports, 77–79, 81
Waste Regulation Authorities, 86, 87, 109
Wastes Inspectorate, 83
Water Quality Inspectorate, 83
West Midlands, 87
Greenpeace, 4, 6, 37, 41, 42, 92, 96, 167, 207, 226, 241
Guinea-Bissau, 226
Gunderson, William C., 217

Haas, Peter M., 13
Haiti, 40, 41
Harbage, Peter T., 217
Harr, Jonathan, 29
hazardous waste. *See* waste, hazardous
Haznews, 72, 173
Haztracks, 213
Heclo, Hugh, 156
Heidenheimer, Arthur J., 156
Héritier, Adrienne, 17, 129, 143, 202
Hong Kong, 214

Ilgen, Thomas, 17, 134, 155
India, 43, 210
Indian Ocean, 40
Inglehart, Ronald, 16
institutions. *See also* individual countries
definition, 2
effectiveness, 12–13
institutional dynamism, 71–72
intermediate-level factors, 51
in international relations theory, 12–15
nested model of environmental regulation, 54
theoretical approach, 2, 17–18, 189–195

integrated pollution control, system of, 82, 230
Integrated Pollution Prevention and Control Directive—1996. *See* European Union
integration, European, 197–199
  bottom-up approach, 198, 201
  conflict/stalemate approach, 199, 201, 202
International Chamber of Commerce, 227
International Council on Metals and the Environment, 227
International Maritime Organization, 209
international versus domestic approaches, 10–12, 18, 65–66
Internet, 168
Ireland, 94
  waste exports to United Kingdom, 78
ISO 14001, 211
Italy, 31, 44
Izmir Protocol, Barcelona Convention, 41

Jacobs, Scott H., 20
Jacobson, Harold K., 14
Japan, 2, 3, 5, 7, 8, 22, 23, 34, 69, 70, 147, 148, 169–182, 194
  Basic Law for Pollution Control, 170
  citizens' movement, 177
  Diet, 170, 173
  Environment Agency, 171, 172, 173, 174, 177, 178
  Environmental Impact Assessment Bill, 182
  government-industry relations, 176–177
  government-society relations, 177–178
  Health and Welfare Ministry, 172
  Hiroshima, 181
  imports, waste, 148, 178–179
  incineration, 173
  Keidanren, 176
  landfills, 173
  Liberal Democratic Party, 169, 172, 173, 181, 239
  Minamata disease, 169, 179
  Ministry for Health and Welfare, 171, 173
  Ministry of Finance, 171
  Ministry of International Trade and Industry, 171, 177
  Nagasaki, 181
  Narita Airport, 177
  Niigata, 179
  prefectures, 171, 172
  Public Center for Waste Treatment, 172
  regulation
    effects on actor behavior/incentives, 176–177
    style, 170, 173–175
  risk assessment, 179
  Tokyo, 171, 172
  waste disposal industry, 179
  waste, hazardous
    definition, 256
  waste management, structure, 171–173
    comparison with Great Britain, 171, 172, 181, 183
  Yokkaichi, 179
  Yokohama, 172
Jasanoff, Sheila, 17, 134, 155
Jordan, Andrew, 133, 203

Kenya, 209
Keohane, Robert O., 13
Khian Sea, 40
Kitschelt, Herbert P., 127, 131, 132
Knill, Christoph, 273
Kohl, Helmut, 123
Kommunekemi, 212
Krasner, Stephen D., 22
Krause, Frank, 129

Lebanon, 214, 243
Lenschow, Andrea, 273

less developed countries, 26, 32, 37, 41, 42, 46, 47, 111
  waste exports to the United Kingdom, 78
Levy, Marc A., 13
Linnerooth, Joanne, 118–119, 125
Lipschutz, Ronnie D., 14
Litfin, Karen, 14
Lomé IV Convention, 41
London Dumping Convention, 36
London Smog of 1952, 81
Love Canal, 29
Luxembourg
  waste exports to the United Kingdom, 78

Maastricht Treaty on West European Union, 200, 202
MacMillan, Harold, 101
mad cow disease. *See* Great Britain, BSE
Mädel, Eyrl, 91
Major, John, 110
Malaysia, 209
maquiladoras, 29, 41
market failures, 33–34
Mauritius, 209
Mediterranean, 41
mercury, 3, 27
Mexico, 37
Moe, Terry M., 17, 107
Montgomery, Mark A., 37, 98, 101
Munton, Don, 33, 217

NAFTA, 37, 41, 213
Namibia, 209
National Association of Waste Disposal Contractors. *See* Great Britain
Netherlands, 143, 229
  National Environmental Policy Plan, 259
  waste exports to the United Kingdom, 78
new social movements, 16
New Zealand, 35

Nigeria, 209
NIMBY. *See* "Not in My Backyard"
Northern Ireland. *See* United Kingdom
"Not in My Backyard," 20, 32, 33, 75, 118, 161, 210, 217
nuclear wastes, 29

OECD. *See* Organization for Economic Cooperation and Development
Organization for Economic Cooperation and Development, 1, 3, 4, 5, 8, 9, 17, 27, 34, 36, 37, 38, 42, 43, 48, 65, 66, 65, 66, 67, 119, 121, 123, 147, 148, 185, 189, 207, 208, 213, 216
  case selection, 69
  data issues (*see* waste, hazardous, data issues)
  decision on transfrontier movements of hazardous waste, 115
  exports of hazardous waste (*see* waste, hazardous, exports)
  imports of hazardous waste (*see* waste, hazardous, imports)
O'Riordan, Timothy, 133

Pakulski, Jan, 166
PCBs. *See* waste, hazardous, types
Pesticides. *See* waste, hazardous, types
Pflügner, Walter, 139
Pharr, Susan J., 170, 177
policy community, 62
policy implementation, environmental, mode of, 63. *See also* individual countries
  flexible systems, 63–64
  rigid systems, 63
political opportunity structure, 63
pollution haven hypothesis, 18–19
Portugal, 45, 230, 241
Postel, Sandra, 103
precautionary principle. *See* individual countries
principal-agent theory, 56

Rabe, Barry, 217
Ramsar Convention, 167
Rappaport, Ann, 172
Raustiala, Kal, 13, 14
ReChem International, 92, 94, 99, 103
regimes
  effectiveness, 13, 14
  environmental, 12, 219
  international, 12, 14, 219, 220, 227
regulation, systems of environmental. *See also* individual countries
  change in, 71–72
  defined, 4
regulatory arbitrage. *See* pollution haven hypothesis
regulatory capture, 231
regulatory convergence, 199–201
  California effect, 200
  Delaware effect, 200
regulatory harmonization, 19, 20, 199, 201, 206
regulatory structure. *See also* individual countries
  allocation of responsibilities among agencies, 58
  defined, 55–56
  federal versus unitary systems, 21–23, 59–60, 195–196, 231
    Australia, 196
    France, 151, 196
    Germany, 196
    Great Britain, 76, 84, 90, 196
    Japan, 175
    United States, 196
regulatory style, 6, 60
  access. *See* access, policy process
  Australia, 70
  conflictual, 60–61
  corporatist, 60–61
  defined, 60
  France, 70
  Germany, 60, 70
  Great Britain, 60, 70
  indicators, 60–61
  Japan, 70

Spain, 60
United States, 60
regulatory systems, 52, 70
Rethmann, Norbert, 145
Rhine River, 123
Rhodes, R.A.W., 82
Romania, 226
Russian Federation, 209

Saudi Arabia, 209
Saunders, Cheryl, 164
Saur, 92
Scharf, Thomas, 129
Schmidt, Manfred G., 121
Schreurs, Miranda A., 14
Schroeder, Gerhard, 138
Scotland. *See* United Kingdom
scrap metals, 43
Seveso incident, 44
Shell Oil, 45, 202
Single European Act of 1987, 93
SITA, 92
Skea, Jim, 90, 200, 202
Skinner, John H., 17
Skolnikoff, Eugene B., 13
Smith, Adrian, 90, 200, 202
South Africa, 209
Southeast Asia, 209
South Pacific, 41
Soviet Union, 37
Spain, 45, 60, 214, 230
  waste exports to United Kingdom, 78
stakeholders, 61
Stockholm Conference on the Global Environment, 11, 81, 123
structure, regulatory. *See* regulatory structure
style, regulation. *See* regulatory style
Superfund. *See* United States
sustainable development, 21
Switzerland, 150
systems, regulatory. *See* regulatory systems

Taiwan, 209
Thatcher, Margaret, 82, 84, 87, 93

Tranter, Bruce, 166
Treaty on West European Union, 93

U.K. Waste Management Ltd., 103
UNEP, 3, 42, 43, 46, 207, 212, 213, 215
United Kingdom, 18, 31, 33. *See also* Great Britain
  Northern Ireland, waste imports, 78
  Scotland, waste exports, imports, 78, 87
  Wales, waste exports, imports, 80, 87
  waste imports by country, 80
United States, 1, 18, 31, 32, 35, 36, 42, 196, 211
  Department of Commerce, International Trade Administration, 172
  Environmental Protection Agency, 84, 86, 162
  Superfund, 31
  USAID, 215

Victor, David G., 13
Vogel, David, 17, 18, 93, 96, 107, 180, 181, 200

Waigani Convention, 41
Wales. *See* United Kingdom
Walloon Waste Case—1992, 228
Waste Directive. *See* DG XI
waste disposal, comparative advantage in. *See* comparative advantage in waste disposal
waste, hazardous
  data issues, 71–72
  defined, 26–27
  disposal technologies, 27–28 (*see also* individual countries)
  exports, OECD countries, 80
  individual OECD countries, 30
  OECD—European countries, 224
  United States, 224
  importation propensity (*see also* individual countries)
  defined, 4, 8
  imports, OECD countries, 80
  incineration, 68, 212, 223 (*see also* individual countries)
  landfills, 68, 223 (*see also* individual countries)
  movements, transfrontier, by individual OECD countries, 38–39
  recycling, 68, 212
  types, 27–28, 78
    PCBs, 3, 78, 103, 105, 154
    pesticides, 78
waste, high risk, 222–223
Waste Management and Research, 72
Waste Management, Inc., 92, 164, 212
waste management industry. *See also* individual countries
  structure, 57–58
waste minimization, 215. *See also* individual countries
waste, special. *See also* individual countries
  European Union regulations, 109–110
waste, toxic
  contrast with hazardous waste, 222
  defined, 222
Weale, Albert, 17, 90, 128, 199, 204, 205
Weidner, Helmut, 176, 180, 182
Weinthal, Erika, 14
Weiss, Edith Brown, 14
Western Europe, 1, 34, 36, 44, 66
Wilson, David C., 105
Woburn, Massachusetts, 29
World Bank, 12, 215
World Health Organization, 215
World Trade Organization, 43
Wright, Vincent, 150
Wynne, Brian, 26, 34, 91

Yakowitz, H., 34, 105
Young, Oran, 13

Zimbabwe, 209

1